Computer Modeling of Water Distribution Systems

AWWA MANUAL M32

Second Edition

American Water Works Association

Science and Technology

AWWA unites the drinking water community by developing and distributing authoritative scientific and technological knowledge. Through its members, AWWA develops industry standards for products and processes that advance public health and safety. AWWA also provides quality improvement programs for water and wastewater utilities.

MANUAL OF WATER SUPPLY PRACTICES—M32, Second Edition
Computer Modeling of Water Distribution Systems

Copyright © 2005 American Water Works Association

All rights reserved. No part of this publication may be reproduced or transmitted in any form or by any means, electronic or mechanical, including photocopy, recording, or any information or retrieval system, except in the form of brief excerpts or quotations for review purposes, without the written permission of the publisher.

Project Manager and Technical Editor: Neal Hyde
Production Editor: Carol Stearns

Library of Congress Cataloging-in-Publication Data.

Computer modeling of water distribution systems.--2nd ed.
    p. com. -- (AWWA manual ; M32)
  Includes bibliographical references.
  ISBN 1-58321-323-6
    1. Water--Distribution. 2. Network analysis (Planning) I. American Water Works Association. II. Series.

TD491.A49 no. M32
[TD481]
628, 1 s--dc22
[628.1'44]   2004058562

Printed in the United States of America

American Water Works Association
6666 West Quincy Avenue
Denver, CO 80235-3098

ISBN 1-58321-345-7

Printed on recycled paper

Contents

List of Figures, v

List of Tables, vii

Foreword, ix

Acknowledgments, xi

Chapter 1 Introduction to Distribution System Modeling, 1

Overview, 1
Historical Development of Distribution System Modeling, 2
Distribution System Modeling Applications, 3
Hydraulic Models, 6
Distribution System Modeling Within the Utility, 9
Surveys, 10
References, 12

Chapter 2 Preparing the Model, 13

Introduction, 13
Model Development Planning, 14
Model Setup, 20
Facilities Data, 21
Demand Data, 27
Operations Data, 35
References, 39

Chapter 3 Hydraulic Tests and Measurements, 41

Introduction, 41
Planning Field Tests, 42
Flow Measurements, 42
Meter Tests, 45
C-Factor Tests, 46
Diurnal Demand Measurements, 48
Pump Tests, 50
Hydraulic Gradient Tests, 51
Fire Flow Tests, 52
Other Tests, 53
Data Quality, 53
References, 54

Chapter 4 Steady-State Simulation, 55

Introduction, 55
Steady-State Calibration, 56
Selecting Limiting Conditions for Design Scenarios, 60
Design Criteria as Analysis Considerations, 67
Developing System Improvements, 74

Continuing Use of the Model, 77
References, 77

Chapter 5 Extended-Period Simulation, 79

Introduction, 79
Input Data, 80
Extended-Period Simulation Setup, 82
Energy Optimization, 90
Case Study City of Fullerton, Calif., 93
References, 95

Chapter 6 Water Quality Modeling, 97

Introduction, 97
Data Requirements, 101
Field Measurements of Water Quality Parameters, 103
Model Calibration, 109
Case Study: North Marin Water District, Calif., 110
References, 116

Chapter 7 Advances and Trends in Network Modeling, 117

Introduction, 117
Automatic Calibration, 117
Tank Modeling, 121
Optimal Control of Pumps in Water Distribution Systems, 125
Summary, 131

Appendix A Hydraulic and Water Quality Modeling of Distribution Systems: What Are the Trends in the US and Canada?, 135

Index, 153

Figures

2-1   Model formulation steps, 20

2-2   Pump curve, 25

2-3   Diurnal curve, 34

3-1   Hand-held pitot gauge, 43

3-2   Traverse positions within a pipe, 44

3-3   Typical velocity profiles at two different flow rates, 45

3-4   Parallel hose method for head loss, 47

3-5   Gauge method for head loss, 48

3-6   Diurnal demand measurement, 49

3-7   Pump tests, 50

3-8   Hydraulic gradient layout, 52

3-9   Hydraulic gradient test, 52

4-1   Flow calibration, 57

4-2   HGL calibration, 57

4-3   Idealized Maximum-day diurnal demand curve, 60

4-4   Equalizing volume requirement, 64

4-5   Equivalent emptying and filling times, 64

4-6   Multiple pump curves, 70

4-7   Pump curve efficiency, 71

4-8   Storage allocation, 74

4-9   Types of storage and elevation, 75

5-1   Residential system demand vs. time, 87

5-2   Golf course system demand vs. time, 87

5-3   Total system demand vs. time, 88

5-4   Example of storage vs. production, Case 1, 89

5-5   Example of storage vs. production, Case 2, 90

5-6   Example of storage vs. production vulnerability analysis, Case 3, 91

5-7   Example of storage vs. production with pumping curtailment, Case 4, 92

5-8   Example of storage vs. production with supplemental power, Case 5, 92

6-1   Set of equations in a typical water quality model, 100

6-2   Protocol for chlorine decay bottle test, 108

6-3   Time series plot comparing modeled results to observed data, 110

6-4   Skeletonized representation of Zone I of the North Marin Water District, 111

6-5	Comparison of observed and modeled sodium concentrations in the North Marin Water District, 113
6-6	Average percent of Stafford Lake water in the North Marin Water District, 114
6-7	Comparison of observed and modeled chlorine residual in the North Marin Water District, 115
7-1	Bi-level computational framework, 119
7-2	Types of tank/reservoir models, 122
7-3	Use of systems model to calculate water age in a reservoir, 123
A-1	Utility responses, 136
A-2	Size of utility, 136
A-3	Population per service connection, 137
A-4	Average per capita demands, 138
A-5	Source water type, 138
A-6	Source water temperature, 139
A-7	Distribution system pipeline length, 140
A-8	Population per mile of pipeline, 140
A-9	Pipe diameters, 141
A-10	Service pressure, 142
A-11	Secondary disinfectants, 143
A-12	Water quality problems, 144
A-13	Size of utility, 145
A-14	Model types, 145
A-15	Current and planned model uses, 146
A-16	Frequency of use, 147
A-17	System demand allocation method, 147
A-18	Types of field measurements, 148
A-19	Information systems tied to model, 148
A-20	Models that are calibrated, 149

Tables

2-1 Water system inventory, 14

2-2 Hazen-Williams C-factors, 23

2-3 Darcy-Weisbach friction factors for pipe roughness in mm, 23

2-4 Operation data required by facility/equipment type, 35

4-1 Typical model scenarios, 61

4-2 Calculation of equivalent emptying and filling times, 63

5-1 System physical parameters for extended-period simulation analysis, 88

A-1 Average day demand, 137

A-2 Pressure zones, 141

A-3 Storage tanks, 142

A-4 Type of problem, 144

A-5 Constituents modeled, 150

A-6 Water quality modeling applications, 152

A-7 Method of water quality calibration, 153

This page intentionally blank.

Foreword

The Computer Assisted Design of Water Systems (CADWS) Committee of the Engineering and Construction Division of AWWA was formed in 1982 and was later renamed the Engineering Computer Applications Committee. The Engineering Computer Applications Committee's mission is to assemble and disseminate information on the use of computer technology in the design, analysis, mapping GIS, and operation of water systems, including use of AM/FM and GIS. The Engineering Computer Applications Committee consists of volunteers, a liaison from the Engineering and Construction Division, and a staff advisor. The committee develops programs for the Annual Conference and specialty conferences, manuals, and other documents.

The purpose of this manual is to compile, discuss, and explain matters relating to the computerized analysis of water distribution system networks for use by engineers, planners, managers, and others involved in the design and operation of water systems. This manual serves as a reference guide and as a manual of practice for those involved in making decisions regarding the implementation and use of distribution system modeling programs. This manual describes a variety of topics that are of primary concern to people involved in water-system analysis of closed-conduit networks or treated-water systems.

This page intentionally blank.

Acknowledgments

The Distribution Modeling Subcommittee was formed to develop this edition of the manual for Computer Modeling of Water Distribution Systems. This group was authorized by the AWWA Engineering Computer Applications Committee of the Engineering and Construction Division under the Technical & Educational Council. The membership of the Distribution Modeling Subcommittee that authored, reviewed, revised, and contributed to this edition included

Susan Ancel, EPCOR Water Services Inc., Edmonton, Alta.
Mark Bolze, Advantica Stoner Inc., Carlisle, Pa.
Nick Braybrooke, Las Vegas Valley Water District, Las Vegas, Nev.
James Cathcart, CGVL Engineers, Lake Forest, Calif.
Lee Cesario, Denver Water, Denver, Colo.
Jeff Cruickshank, Hazen & Sawyer, P.C., Greensboro, N.C.
Walter Grayman, W.M. Grayman Consulting, Cincinnati, Ohio
Ron Hinthorn, Advantica, Carlisle, Pa.
Richard Humpherys, Carollo Engineers Inc., Phoenix, Ariz.
Amy Purves, Plangraphics Inc., Silver Spring, Md.
Lewis Rossman, USEPA, Cincinnati, Ohio
James Thomte, PB Water, Albuquerque, N.M.

Additional contributions and reviews by

Thomas Walski, Haestad Methods Inc., Nanticoke, Pa.
Lindell Ormsbee, University of Kentucky, Lexington, Ky.
Robert Mahoney, Brown and Caldwell, Lakewood, Colo.

Special thanks to Richard Humpherys for his significant efforts in resolving a number of issues with the drafts from each of the committee members and developing a cohesive document to move forward in the review process.

The membership of the Engineering Computer Applications Committee at the time this manual was approved was as follows:

Susan Ancel, Chair, EPCOR Water Services Inc., Edmonton, Alta.
J.L. Anderson, CH2M Hill, Louisville, Ky.
William Archambault, J.R. Holzmacher, Hauppauge, N.Y.
Dominic Bocelli, CDC, University of Cincinnati, Cincinnati, Ohio
Stephen Bopple, Ohio Environmental Protection Agency, Twinsburg, Ohio
Lawrence Catalano, Farnsworth Group, Denver, Colo.
J.A. Cathcart, CGVL Engineers, Lake Forest, Calif.
Lee Cesario, Denver Water, Denver, Colo.
Jeff Cruickshank, Hazen & Sawyer, P.C., Greensboro, N.C.
David Dennis, City of Cleveland Division of Water, Cleveland, Ohio
Jeff Dieterlin, Veolia Water Indianapolis, LLC, Indianapolis, Ind.
C.B. Gale, HNTB Corporation, Indianapolis, Ind.
Walter Grayman, W.M. Grayman Consulting, Cincinnati, Ohio

A.M. Green, Advantica, Loughborough, UK
Richard Head, Philadelphia, Pa.
R.T. Hinthorn, Advantica, Carlisle, Pa.
C.D. Howard, Victoria, B.C.
P.H. Hsiung, MWH Soft Inc., Shawnee Mission, Kan.
R.A. Humpherys, Carollo Engineers Inc., Phoenix, Ariz.
Laura Jacobsen, Las Vegas Valley Water District, Las Vegas, Nev.
K.T. Laptos, Gannett Fleming Inc., Harrisburg, Pa.
M.V. Lowry, Turner Collie & Braden Inc., Austin, Texas
Norman Nelson, Bristol-Myers Squibb Company, Princeton, N.J.
Bill Orne, Engineering Resources Corporation, Chapin, S.C.
C.M. Rethamel, Farnsworth Group, Denver, Colo.
Houjung Rhe, Haestad Methods Inc., Morgan Hill, Calif.
E.P. Skipper, Newport News, Newport News, Va.
Jerald Stevens, Waterloo Water Works, Waterloo, Iowa
Adam Strafaci, Haestad Methods Inc., Waterbury, Conn.
J.C. Thomte, PB Water, Albuquerque, N.M.
Bill Lauer, staff, American Water Works Association, Denver, Colo.

AWWA MANUAL M32

Chapter 1

Introduction to Distribution System Modeling

OVERVIEW

Water utilities seek to provide customers with a reliable, continuous supply of high-quality water while minimizing costs. This water is often delivered through very large and complex distribution systems consisting of many miles of pipe and often containing numerous pumps, regulating valves, and storage reservoirs. These systems are often difficult to understand because of their physical complexity, and because of the large amount of data that must be processed. Sometimes, key pieces of information needed to understand a system are not available. In addition, the chemical interactions that take place in the water, and between the water and pipes or reservoirs, are complex. One tool that has evolved over time to help water system designers, operators, and managers in their task of delivering safe, reliable water at a low cost is distribution system modeling.

Distribution system modeling involves using a computer model of a water distribution system to predict the behavior of the system to solve a wide variety of design, operational, and water quality problems. The computer model is used to predict pressures and flows within a water distribution system to evaluate a design and to compare system performance against design standards. The model is used in operational studies to solve problems, such as evaluating storage capacity, investigating control schemes, and finding ways to deliver water under difficult operating scenarios. Water quality models are used to perform such tasks as computing water age, tracking chlorine residuals, and reducing disinfection by-products in a distribution system.

Distribution system modeling began with the advent of analog computers and has evolved over time as computer software and hardware developed to become more powerful and easy to use. Models containing thousands of pipes are created and used on readily available personal computers. Models that once took hours to run are now run in seconds or fractions of a second. Originally, models were used only to solve for pressures and flows within a distribution system. Although this capability remains at the very core of all water distribution modeling work, hydraulic models or their associated modules can now calculate water age, energy costs, perform isolation analyses, and do optimization of various modeling parameters.

Historically, model building was a very expensive and labor-intensive task. Now that models can effectively share data with Geographic Information Systems (GIS), Supervisory Control and Data Acquisition systems (SCADA), and Customer Information Systems (CIS), the effort to create and maintain a model is reduced. Information obtained from a model study is filtered, organized, and presented in a variety of graphical and nongraphical ways so that the results of a study are more easily understood by a nonspecialist. These advances in technology have broadened the uses of distribution system modeling from just an infrastructure-planning tool to a system used to improve operations, analyze water quality, and to plan water system security improvements.

This manual was developed by the Engineering Computer Applications committee of the American Waterworks Association (AWWA). The purpose of this manual is to share our collective expertise on distribution system modeling so that it is understood and applied more effectively, to the benefit of water utilities and water consumers everywhere.

HISTORICAL DEVELOPMENT OF DISTRIBUTION SYSTEM MODELING

Pre-1970s

Manual calculations for small-pipe systems were used through the 1960s. The Hardy-Cross method was sufficient for single-loop systems, but without the aid of a computer it was impractical for systems having several loops. An analog computer model was created using electronic circuitry in the 1950s. These were large, physical models that were expensive and difficult to use. Mathematical computer models appeared in the 1960s. These were simple models that used system data files and system solution algorithms to solve for pressure and flow in the system.

1970s Through 1980s

Software packages were sold with a variety of features. Steady-state and extended-period simulation models became standard features. Graphics were used for drawing the system and displaying output. Software was appended with the term *packages* because it contained several modular components that were compatible. Some software packages utilized other specialized software for data entry, display, and reporting of results.

1990s

The 1990s experienced exponential growth of system modeling capabilities. EPANET was a modeling program developed by the US Environmental Protection Agency (USEPA) to support ongoing research and made available to the public. Some vendors have taken the EPANET model and added an improved user interface. Software packages were designed to be compatible with other standard software packages, such as Microsoft® Word and Excel, AutoCAD, databases, and GIS software for editing, database, and drawing functions. The result was a familiar user interface and the ability to utilize existing software rather than having to create and update new software. Water quality extended-period simulation became standard features within system modeling software packages.

Present

Software packages are more effectively working with, and sometimes as part of, GIS software in response to the dependence on GIS systems by water utilities. GIS data are becoming more common, and the quality of the GIS data is improving, significantly reducing the effort required to develop models. Optimization tools are available for use in optimizing design, as well as in aiding the calibration process. Recent security concerns have resulted in studies to develop emergency response plans and to evaluate the impact that various disasters may have on the water distribution systems. Models are used increasingly for water quality analyses, such as evaluating water age and disinfection by-products. USEPA has also allowed hydraulic modeling as a means of determining preferred locations for water quality monitoring sites that are required to meet regulatory requirements.

DISTRIBUTION SYSTEM MODELING APPLICATIONS

Benefits of Computer Modeling

Before the use of computerized models, water distribution system analysis using hand-held calculators and slide rules involved many simplifying assumptions and approximations. As a result, designs were often much more conservative and expensive than necessary. To solve hydraulic system problems, there must be one equation for each pipe, pump, and valve, or for each junction, depending on the method used to solve for the unknowns in the hydraulic calculations. The number of equations that must be set up and solved in a system hydraulics problem is very large, even for the most basic water distribution system. The value of a computer model is that tedious calculations are performed very fast and more accurately than by manual calculation. In addition, the computer is an effective means of managing the large amounts of data necessary to analyze a water distribution system. By using computer models, decision makers can focus on formulating and comparing alternatives as well as communicating results, rather than on the procedural mechanics of solving system equations.

Computer models of water distribution systems are not an end in themselves but are a tool to help managers, engineers, planners, and operations staff. Their real purpose is to support decision-making processes in planning, design, and operation of water distribution systems. When properly implemented, the model is an integral part of the utility's decision-making process. Engineers and operators of a water system are still ultimately responsible for decisions based on output from computer models.

Distribution system modeling software generally falls into four application categories: planning, engineering design, systems operations, and water quality improvement.

Planning

A primary planning application of distribution system analysis software is in assisting development of long-range capital improvement plans, which include scheduling, staging, sizing, and establishing preliminary routing and location of future facilities. Other applications include the development of main rehabilitation plans and system improvement plans. Rehabilitation plans identify and prioritize mains that need to be cleaned and/or lined. Distribution system improvement plans identify where installation of new mains, storage facilities, and pump stations are necessary to keep pace with growth or new utility standards. The following are examples of specific system analysis planning applications.

Capital Improvement Program. Water utilities usually have a master plan that identifies capital improvements. A model is usually used to identify these capital improvements to respond to projected growth or to replace aging infrastructure.

Conservation (Impact of) Studies. Water conservation is used to stretch limited water supplies or to reduce water use so that capital improvements are delayed or eliminated. A model is useful to apply projected demands with conservation measures to evaluate their potential for success.

Main Rehabilitation Program. A model is used to identify the effect of specific mains that are bottlenecks in the system either because demands have increased significantly or because of encrustation. The model is used to determine the hydraulic effect of rehabilitating the main to evaluate the effectiveness of rehabilitation alternatives.

Reservoir Siting. A reservoir should be located where there is good turnover in the reservoir, where the reservoir effectively meets peak demands, and is filled during off-peak demand times. The model is used to explore these scenarios to fine-tune preferred hydraulic solutions.

Engineering Design

Engineering design applications include the sizing of various types of facilities. Pipelines, pump stations, pressure regulating valves, tanks, and reservoirs are sized using pressures and flows that result from distribution system modeling. In addition, system performance is analyzed under fire flow conditions and adjustments are made to meet fire demand. Following are examples of engineering design problems that are solved using computer models.

Fire Flow Studies. The model is used to simulate fire flow demands at hydrant locations throughout a city to determine how much water is delivered at fire hydrants. Where deficiencies are discovered, the distribution system is improved with main reinforcements or looping. These studies are also used to demonstrate compliance with fire protection standards.

Valve Sizing. A distribution system often has pressure-regulating or pressure-sustaining valves to direct the flow to a different zone. Distribution systems may also have throttle valves to direct flow within a zone to different reservoirs or storage locations. The model is used to determine how much flow is required through these valves so that the valves are sized appropriately.

Reservoir Sizing. Reservoirs are often sized by estimating the total diurnal flow, fire flow, and emergency storage requirements within a particular zone. However, reservoir capacity should also consider the rate of water delivery to the reservoir location and the size of the distribution area. A model is useful to evaluate inflows and outflows to a reservoir in order to determine an optimal size for a particular location or to specify other improvements so that a reservoir at the preferred site is adequately served by transmission mains and pumping stations.

Pump Station/Pump Sizing. Models are used to calculate system curves of distribution systems so that pumps are selected that provide the necessary head and flow. Proposed pumps are then used in the model across the range of operating conditions to determine how well they meet a variety of operating conditions.

Calculation of Pressure and Flow at Particular Locations. A water distribution system must provide adequate amounts of water at pressures that are within a range specified by the standards used by the water utility. A model's core functionality is pressure and flow hydraulic calculations. Models are used to predict pressures under specific demand conditions under a wide variety of scenarios to identify low pressures and to select the infrastructure that will remove flow or pressure deficiencies.

Zone Boundary Selection. Most water distribution systems deliver water to customers located at a range of different elevations. Distribution systems are separated into pressure zones that follow consistent elevation contours in order to keep pressures within reasonable ranges. Models are useful to evaluate potential zone boundaries and to determine the adequacy of infrastructure that delivers water to each zone.

Systems Operations

Applications for operations include assisting in the development of operating strategies, operator training programs, and system troubleshooting guidelines. Operating strategies may be driven by emergency conditions, energy management, water availability, etc. For example, contingency plans are developed in the event a key facility, such as a pump station, fails. Distribution system modeling is also used to develop operational strategies for energy management and water quality guidelines. Strategies for shifting supply between treatment plants are developed that determine efficient use of available water; optimizing these strategies results in efficient use of pipeline capacities, tank levels, and required treatment plant production, among other things.

Personnel Training. Models are used for training personnel that operate the distribution system. System operators can experiment with the model to determine how the system responds to changes in operating conditions.

Troubleshooting. Models are used to troubleshoot the cause of various problems, such as low pressure, water circulation problems, and events that would otherwise be inexplicable.

Water Loss Calculations. In the event of a major main break, the model is used to estimate the amount of water lost through the break as may be required for damage assessments.

Emergency Operations Scenarios. Water distribution systems often have critical components; if the components fail, water delivery is interrupted. A model is useful to evaluate the potential impact of a failure and to devise means of reducing the damage or impact of a critical component failure.

Load Shifting Between Treatment Plant Studies. Water treatment plants are sometimes taken out of service for repairs or because the water supply is unavailable for a time. Furthermore, the quality of water at one source may be better at certain times of the year, so the use of the high-quality source is maximized. The model is useful to devise operating scenarios to utilize water sources to achieve desired objectives.

Model Calibration. Model calibration is typically thought of as a step in developing a useful model. However, the calibration process is useful to operations staff in discovering anomalies in the distribution system, such as closed valves, tuberculated pipes, leaks, or incorrect infrastructure data. This information, once discovered through the calibration process, can explain operational difficulties, and identify distribution system problems that will improve the operation of the system when resolved.

Main Flushing Programs. A hydraulic model is an excellent tool to develop a main flushing program. The model is useful to identify flow paths in the distribution system so that the flushing locations and sequence are established.

Area Isolation. Water utilities frequently need to isolate an area for maintenance or other work. Often, it is helpful to identify those customers whose service will be interrupted by the isolation event. In addition, those planning the event need to know which valves to close in order to minimize the impact of the isolation. Hydraulic models are a tool used to accomplish this task.

Water Quality Improvements

Water quality regulations in the United States are regulating the level of disinfection by-products (DBPs) in a water distribution system. Standards and expectations for water quality have increased the demand for water quality analysis in the distribution system. Following are examples of how distribution system modeling is used to improve water quality.

Substance Tracking. If a contaminant enters the distribution system through a treatment plant, well, reservoir, or other location, it will spread through the distribution system, affecting the water quality of consumers who receive water from that source. The contaminated water may also mix with water from other sources. A model is useful to predict the contaminant level and zone of influence of the contaminant in the distribution system. Customers who are affected by the contaminant are identified, and portions of the distribution system that need to be flushed are also defined.

Water Source/Age Tracking. Water age is an important water quality parameter in a distribution system. Chlorine levels decay over time, and DBP levels tend to increase with time as the chlorine reacts with organic compounds in the water. To maintain water quality, water utilities are striving to minimize water age. This is done by ensuring that water in reservoirs turns over regularly, so that it does not become stagnant, and by minimizing dead ends. When multiple water sources serve an area, operating strategies are devised to reduce water age where possible. Distribution system modeling helps identify operating strategies to reduce water age.

Chlorine Levels. A model is used to predict chlorine decay in a distribution system. This is useful to determine chlorination levels at the treatment plant and to select rechlorination sites where necessary to boost chlorine levels.

Water Quality Monitoring Locations. Part of the water quality regulations proposed by USEPA include selecting appropriate sites to place permanent water quality monitors to monitor DBP levels and demonstrate compliance with federal regulations.

HYDRAULIC MODELS

A computer model is composed of two parts: a database and a computer program. The database contains information that describes the infrastructure, demands, and operational characteristics of the system. The computer program solves a set of energy, continuity, transport, or optimization equations to solve for pressures flows, tank levels, valve position, pump status, water age, or water chemical concentrations. The computer program also aids in creating and maintaining the database, and presents model results in graphical and tabular forms.

Model Data

Model data is usually associated with two entities: links and nodes. Links represent pipes, pumps, valves, and sometimes tanks in a distribution system. Nodes represent junctions, endpoints, and locations of elevation extremes, water sources, demand points, and sometimes tanks. The characteristics of facilities, such as pipes, include length, roughness or friction coefficient, diameter, and reaction coefficient. Operational parameters are also associated with the respective links. Attributes of nodes include elevation, grade line, water demand, and water supply. Model databases include metadata and descriptive information that is useful in defining, organizing, and managing the model.

The model is a valuable asset to the user and represents a substantial effort in data collection, entry, and quality control. Therefore, this investment in data and in an understanding of the distribution system should be protected. One of the keys to maintaining the value of a hydraulic model is making sure that the model is updated with new infrastructure, demands, and operating information. In order to maintain confidence in the results of the model, data within the model should represent the current system configuration so that stakeholders will trust and use the model.

Modeling Software

The heart of the water distribution system modeling software is a system solution algorithm, which sets up and solves the hydraulic and transport equations. Depending on the problem, there are several kinds of equations that are solved.

Continuity Equations keep track of water flow, make certain that the total flow into a node equals the flow out plus or minus any changes in storage or demand.

Energy Equations account for energy loss caused by friction in pipes, valves, and fittings, and energy gained in pumps (the total energy change around a loop is zero).

Transport Equations in water quality models account for the movement of substances (pollutants or tracers) through a distribution system and any reactions that may occur.

Cost Equations in optimization models account for energy costs or cost of piping.

In addition to a computation engine, the software must manage the databases or files containing model information, interact with the user, and present model results in both graphical and nongraphical forms. The following list of features are commonly available in distribution system modeling software packages, and should be included in any modeling package for use by a water utility.

Steady-State Analyses. Packages will have the ability to perform a steady-state analysis, which takes a "snapshot" of pipe system conditions at any instant in time. Steady-state analyses are typically used to evaluate maximum day, peak hour, and fire flow conditions.

Extended-Period Simulation. A package should also provide the ability to perform a sequence of analyses with the output from each analysis forming the input to the next analysis. This capability may be used, for instance, to model the operation of a water system over a 24-hour period with an analysis run each hour. Such a simulation is useful in modeling variations in demand, reservoir operations, water quality, and water transfers through transmission pipelines. Extended-period simulation requires that the system package model flow and pressure switches, incorporate demand hydrographs for nodes, and allow for varying tank configurations.

Graphical User Interface. A package should provide a graphical user interface (GUI), which allows the user to see a schematic of the distribution system displayed on the monitor screen. Node and link data are displayed by clicking on the entity on the screen, and model results are also displayed for each entity. Modeling results are presented graphically, in addition to tabular form, giving the modeler a better understanding of distribution of flows and pressures throughout the system. GUIs can be either proprietary (written specifically for the package), or they can be customized and generated from standard graphics packages, including computer-aided design and drafting (CADD or CAD) packages.

Error Reporting. A package should test for distributions system configuration errors. For example, any portions of the distribution system that are not connected to the rest of the system should result in an error message indicating the location where the model is discontinuous.

Selective Reporting of Results. The user is able to specify the results to be reported in tabular form, so pages of output need not be generated after each run, as the user may only be interested in results in one specific area of the system. This user-specific reporting saves hard disk space and paper while speeding the user's review time.

System Components. As well as pipes and nodes, a package should have the ability to model pumping units (incorporating pump characteristic curves) as well as flow- and pressure-regulating devices.

Data Management. Modelers should be able to export and import model data and model results to and from other applications, such as spreadsheets, databases, and GIS systems. These capabilities are widely available and an important part of any modeling package.

Automated Fire Flow Calculation. Some distribution system modeling packages automatically calculate the available fire flow at each node. These calculations are useful in identifying areas having weak firefighting capability.

Scenario Generation. Distribution systems with any level of complexity are modeled more easily by applying various combinations of demands, facilities, and operating parameters (such as regulating valve and pumping unit settings). System modeling packages may allow variation and combination of these three types of data in a simulation by keeping them in separate databases for specific or combined model access.

Water Quality. Utilities are increasingly interested in modeling the water quality within a distribution system, particularly the decay of chlorine residual and water age. The ability to perform water quality analysis should be a standard part of any modeling package.

Hardware

Typical hardware for modeling consists of a personal computer, printer, and plotter. The computer should meet the modeling software requirements. Color plotting or printing is helpful to communicate model results. If the organization is large enough to have a computer network and a network server, the modeling system should utilize the server for data backups and storage. Regular data backups are required. The software vendor should be consulted for specific hardware recommendations at the time software is purchased.

Related Software Systems

Information management trends within utilities are moving toward better information sharing so that decision makers can have as many information resources to make the best decisions. This is often done by using both common databases and files that are shared by a variety of software applications. Distribution system modeling uses a wide variety of information about physical assets, customers, billing data, geographical information, and operational information. Furthermore, modeling activities can benefit a variety of groups within the utility, strengthening the need to communicate and share information. Brief descriptions of software systems, information systems, or corporate-wide databases that are in some way related to distribution system modeling are listed below.

Geographical Information System (GIS). A GIS stores and displays information that is best understood in a geographical context, i.e., information that is easily understood by displaying on maps. The spatial relationships between entities is significant for most information that is stored in a GIS. A GIS has the potential to store vast amounts of information that is useful for system analysis. This data can include pipe assets, customer meter locations, zoning and land parcel data, aerial photographs or other land bases, street locations, digital terrain models (DTMs), and jurisdictional boundaries.

Information in a GIS is saved in file formats that some modeling software packages can read. Alternatively, information in a GIS database is translated into a form that is imported into the model database. Some GIS land-base information and other data layers are displayed directly within some modeling packages. The usefulness of GIS pipe data is often dependent on the way that the information is collected and stored in the GIS database. Pipe data in the GIS is most useful if the topology, or connectivity, is already established in the GIS.

Computer-Aided Design and Drafting (CADD). CADD systems are used to create maps of water distribution systems. Therefore, they are a source of pipe information that can be transferred to the model. In addition, CADD systems are a useful means of displaying model information and results. Some modeling packages bring CADD information into the model using .DXF files.

Supervisory Control and Data Acquisition (SCADA). SCADA systems are used to remotely control the operation of pump stations, valves, and other system infrastructure. They are also useful to collect data that includes pressures, flows, reservoir levels, valve positions, pump status and speed, chlorine levels, and other information that is useful to monitor the system. This information is collected at frequent intervals and stored for extended periods of time. SCADA is a good source of operational information, as well as calibration data. SCADA data is also used to define the starting point for operational analyses by taking the data and using it to define the boundary conditions that are placed in the model. SCADA data usually does not go into the model directly. An interface is usually required that could be as complex as a custom software routine, or as simple as importing SCADA data into a spreadsheet via a CSV file and formatting the data to import it into the model.

Customer Information System (CIS). A CIS is useful to provide customer water-use information to develop demands. Typically, average annual water usage and customer rate class for each customer is extracted from a CIS; then GIS, modeling, or customized tools are used to link these demands to nodes in the model. The specifics of how the CIS data are linked and entered into the model is highly dependent on the data and software used at the utility.

DISTRIBUTION SYSTEM MODELING WITHIN THE UTILITY

A successful distribution system modeling program functions best with a team of individuals who fill all the roles required for a system analysis program and are able to effectively provide system modeling results to decision makers within the utility. Issues that often need to be addressed when implementing a modeling program are outlined in this section.

In-House vs. Outside Consultants. The utility should decide whether the model is developed and maintained by the utility or by an outside consultant.

Usually, a utility understands its system very well and has easy access to model-related information. However, a utility may not have the expertise or resources to develop and maintain the model. Some utilities construct and run their own models, while others hire outside consultants to perform some or all of the modeling work. A model owner who is committed to maintaining the model and to developing modeling expertise is essential for an in-house modeling program. If consultants do the modeling, ownership rights of the model and data should be clearly delineated in a contract. Regardless of who does the modeling work, a long-term commitment to maintaining the model and to having experts utilize the model regularly are necessary components of success.

One-Time vs. Long-Term Use. Many decisions made during model development depend on whether the model is used for solving a short-lived problem or for periodic use over an extended period of time. If the model is used for a specific problem, questions regarding the level of detail are easily answered based on satisfying the needs of the problem. If the model is to be used for many purposes, the model should be developed to serve the most demanding applications and simplified, if necessary, for other applications. For example, a model developed to assist in master planning may not contain enough detail for determining available fire flows in subdivisions or for water quality modeling. Decisions must be made on whether this level of detail should be included in the model or added later, if required. The power of today's computers make it practical to use a complex model for even a simple task. However, many software packages have tools to simplify a model when necessary.

Model Developer vs. Decision Maker. There are two distinct roles in model development: the role of the modeler and the role of the decision maker. The modeler is the person responsible for initial development and running the model. The decision maker is a professional engineer or licensed operator who interprets and makes decisions based on outputs from the model. These two roles could be filled by one individual, two individuals, or even two different companies. The key element is that the decision maker must be satisfied that the modeler has indeed developed a system model that is adequate for the problem or problems being considered. Calibration and sensitivity analyses are two methods available to the decision maker to assure that the model is adequate for the intended purpose.

Modelers vs. Rest of Utility. It is essential that modeling, whether performed in-house or by a consultant, be done with the awareness and cooperation of the rest of the utility. While some individuals may serve as the experts on the model, all interested parties should have input in model development, understand the capabilities and limitations of the model, and appreciate the important role of modeling in decision making. Modeling should be done with thorough consideration of utility operations. For example, utility operators have great insights into the operation of a system, as well as its physical limitations. By working with the operations staff, the modeler can incorporate operators' insights into the model, and operations staff can become sufficiently comfortable with the model to trust its results.

SURVEYS

Since 1980, the AWWA Engineering Computer Applications Committee and other groups have surveyed water utilities to compile information about the use of system modeling packages. In the 1999 survey, a total of 989 utilities were selected from the AWWA database based on serving populations greater than 35,000 people. A total of 174 water utilities responded to the survey; 150 of these utilities were using a distribution system modeling package. The paper, Hydraulic and Water Quality Modeling of Distribution Systems: What Are the Trends in the U.S. and Canada? discusses the results of the survey and is presented in appendix A. Notable results of the survey follow:

- Although a total of 20 different distribution system modeling packages were used by the respondents, the market was dominated by a relatively small number of packages. System modeling packages are typically used weekly or monthly.
- Seventy-seven percent of respondents described the accuracy of their model (accuracy of 60–100 percent) 97 percent gave no response at all.
- Most respondents used their model for steady-state analysis, with only a few performing extended-time simulations.
- Billing records were typically used to generate node demands.
- Few utilities linked their system modeling packages to other software, but many had plans for multiple links in the future.

Trends

Several significant trends in system modeling have become apparent through the 1999 Network Modeling Survey, presentations and discussions at conferences, and *Journal AWWA* articles. Some trends are now well established while others are still in their infancy.

Use of Personal Computers. Personal computers (PCs) now comprise the dominant hardware used for system modeling. Some large utilities maintain models on networked workstations, allowing simultaneous access to a centralized database. However, PC-based system modeling packages easily model distribution systems containing thousands of pipes in a matter of seconds.

Graphical User Interfaces. Distribution system modeling packages are capable of graphical input and output, easing data entry, and evaluation of results.

Water Quality Analyses. Many system modeling packages have the ability to model water quality parameters in a pipe network and in reservoirs, which utilities find valuable in response to new water quality regulations and to heightened public interest in water quality.

Common Databases. In some utilities, computer systems are organized around a type of architecture where common databases are shared by many applications. In such a framework, a distribution system modeling package extracts data from a large *enterprise* database that is shared by many work groups.

Model Sophistication. Model surveys reveal a strong trend toward all-main models and extended-period simulation to examine water quality issues in distribution systems.

Demand Allocation. Although no new demand methodologies were identified in the latest survey, there was greater emphasis on using GIS tools in model loading.

Frequency of Use. Utilities intend to use water models more frequently in the future. Specific uses listed include master planning, fire flow evaluation, and energy and operations optimization.

Information Systems Integration. The near term will see a far greater integration of information systems such as GIS, SCADA, Customer Information Systems, and CAD.

Real-Time Modeling. In the past, distribution system modeling packages were typically too slow and unwieldy for system operators to use in generating operating strategies and testing "what-if" scenarios. High-speed processing and data input available from SCADA allow utilities to provide modeling capabilities to their operators. Careful consideration must be given to the user interface in this regard, and simplified models may be required.

REFERENCES

Anderson, Jerry L., Lowry, Mark V., Thomte, James C. 2001. Hydraulic and Water Quality Modeling of Distribution Systems: What Are the Trends in the U.S. and Canada? In *Proceedings of the AWWA Annual Conference.* Denver, Colo.: American Water Works Association (AWWA).

Benedict, R.P. 1910. *Fundamentals of Pipe Flow.* John Wiley & Sons, New York.

Bhave, P.R. 1991. *Analysis of Flow in Water Distribution Systems.* Technomics Publ., Lancaster, Pa.

Casey, Rob. 2001. Integrating Modeling, SCADA, GIS, and Customer Systems to Improve Network Management. In *Proceedings of the AWWA Information Management and Technology Conference.* Denver, Colo.: AWWA.

Eggener, C.L. and L. Polkowski. 1976. Network Modeling and the Impact of Modeling Assumptions. *Jour. AWWA* 68:4:189–196.

Gessler, J. 1980. Analysis of Pipe Networks. In *Closed Conduit Flow.* Chaudry, M.H. and V. Yevjevich. Water Resources Publications, Littleton, Colo.

Gupta, R. and P. Bhave. 1996. Comparison of Methods for Predicting Deficient Network Performance. *Journal of Water Resources Planning and Management* 122:3:214.

Haestad Methods, Inc. 1997. *Computer Applications in Hydraulic Engineering.* Haestad Methods, Inc., Waterbury, Conn.

Hauser, B.A. 1993. *Hydraulics for Operators.* Lewis Publishers, Ann Arbor, Mich.

Jeppson, T.W. 1976. *Analysis of Flow in Pipe Networks.* Ann Arbor Science Publishers, Ann Arbor, Mich.

Mays, L.W. 1999. *Water Distribution Handbook.* McGraw-Hill, New York

Mays, L.W. 1999. *Hydraulic Design Handbook.* McGraw-Hill, New York

Nayar, M.L. 1992. *Piping Handbook.* McGraw-Hill, New York

Seidler, M. 1982. Obtaining an Analytical Grasp of Water Distribution Systems. *Jour. AWWA* 74:12: 628–630

Stephenson, D. 1976. *Pipeline Design for Water Engineers.* Elsevier Scientific Pub. Co., Amsterdam, New York

Walski, T.M. 1984. *Analysis of Water Distribution Systems.* VanNostrand Reinhold, New York.

Walski, T.M., Gessler, J., and J.W. Sjostrom. 1990. *Water Distribution—Simulation and Sizing.* Lewis Publishers, Ann Arbor, Mich.

Williams, G.S. and A. Hazen. 1920. *Hydraulic Tables.* John Wiley & Sons, Inc., New York.

Wood, D.J. and C.O.A. Charles. 1972. Hydraulic Analysis Using Linear Theory. *Journal of the Hydraulics Division. American Society of Civil Engineers* (ASCE), 98:7:1157–1170.

Wood, D.J. and A.G. Rayes. 1981. Reliability of Algorithms for Pipe Network Analysis. *Journal of the Hydraulics Division. ASCE*, 107:10:1247–1248.

WRc Plc. 1989. *Network Analysis—A Code of Practice.* Water Research Centre, Swindon, United Kingdom.

AWWA MANUAL M32

Chapter 2

Preparing the Model

INTRODUCTION

One of the major tasks of distribution system modeling is developing and maintaining a model that represents the physical network with a degree of accuracy sufficient for the utility to have confidence in it. Information collected to create a model comes from a variety of sources. The type and quality of data available at each utility is unique. Therefore, the modeler responsible for developing a hydraulic model must use creativity in exploring sources of data collection. The modeler should also use good engineering judgment to evaluate data quality and assess the impact of data quality on model results.

Distribution system models should be created and maintained with the end goal in mind so that the standards of accuracy, completeness, and level of detail are appropriate for the modeling application. The utility's best interest is to develop a quality model without wasting the effort to collect information at a level of precision that fails to provide additional benefits. The model development work begins with careful planning to define modeling requirements and to investigate the sources of information that are available to develop the model.

Data used to create a model are classified as geographical information, facilities data, operational data, and demand data. Geographical data include a land-base aligned to the desired geographical coordinate system, jurisdictional boundaries, street centerlines, aerial photographs, and other information useful to establish the physical location of the model. Facilities data include all physical descriptions and attributes of the pipes, pumps, valves, and reservoirs. Operating data includes parameters such as flow rates, reservoir elevations, pressure-reducing valve setpoints, valve or pump controls, valve or pump status, and fixed pressures that form boundary conditions in the model. Demands are water consumption values assigned to nodes throughout the model. Demands are adjusted to represent a wide variety of conditions. For extended-period analysis, demands are set to change with time. Some types of analyses will require additional information, depending on the type of study. Energy costs, infrastructure costs, water constituent or contaminant concentrations, and demand or customer growth projections are examples of additional data that may be required.

MODEL DEVELOPMENT PLANNING

Before developing a distribution system model, the modeler determines the purpose or objective of the analysis. The scenarios to be assessed dictates the type, quantity and required accuracy of the data collected. A review of available data sources also assists in determining the effort and resources required for data collection, data entry, and model calibration.

Characterizing the System

Each water distribution system is unique—the modeler must consider system structure and requirements in developing the hydraulic model. Data collected falls into four main categories: (1) geographical data, (2) facilities data, (3) operational data, and (4) demand data. A detailed water system inventory reveals much of this information (see Table 2-1). However, accumulating abundant data, in itself, is not sufficient. It is important to incorporate the data into the model in a manner that reflects both the actual physical components of the system and the actual operating conditions. Judgments should be made to resolve data conflicts and ambiguities.

Table 2-1 Water system inventory

Identify and collect geographical data:

- Land base
- Aerial photographs
- DTM, GPS, contour map, or other elevation data
- Street maps

Establish inventory of existing facilities:

- Pipe locations, diameter, length, type, age, and condition
- Storage facilities, including capacity, location, availability, dimensions, water surface elevations, and system connections
- The design and operating features of service and booster pump stations, including capacity, system head, elevation, and related features
- Elevation, operability, and pressure settings on control valves and regulating valves
- Service area and pressure zone boundaries

Review system operating records and interview staff:

- Diurnal and seasonal treated water production records
- Service and booster pump flows and pressures
- Clearwell and storage level variations for various demand scenarios
- System operation criteria and setpoints for pump stations, storage, and service pumps
- System operating pressures determined from previous testing, monitoring, and fire flow testing and rating
- Power consumption records, costs, and rate structures
- Observed low-pressure areas indicating system deficiencies
- Observed fire flow capabilities

(Table continued on next page)

Table 2-1 Water system inventory (continued)

Review water consumption records:

- Total yearly water sales (billing records)
- Service population
- Number and type of service connections
- Consumption by meter route and by major users
- Peak factors
- System diurnal demand curve (A diurnal demand curve is a dimensionless profile that describes how demands change in a 24-hour period. The diurnal profile value at a particular hour is multiplied by the average daily demand to obtain an hourly demand.)

Develop water demands using existing water consumption data, and population and land use projections. Establish current and projected water demands for the total system, sub-areas within the system, and major users on the following basis:

- Average annual day demand (Total water usage in a year divided by the days in a year.)
- Maximum day demand (The highest demand occurring in a single day during an entire year.)
- Peak hour demand (The highest hourly demand that would occur in a single year.)
- Fire flow demand
- Minimum hour demand or maximum storage replenishment rate
- Other critical demands within the utility's service area
- Land use classifications
- CIS databases with customer addresses and water use records.

Establish background information for power management and operational improvement programs:

- Determine the average maximum day pumping rate from the developed diurnal demand curve
- Determine whether or not system storage or auxiliary power can be used to reduce the peak pumping rate to a value less than or equal to the average maximum day demand
- Check the piping arrangement to storage complexes for replenishment constraints
- Plot pump characteristic curves versus system head conditions
- Examine pump curves versus system head conditions to determine optimum operation sequence based on system demand

Sources of Network Data

Traditional sources of distribution system data were paper records maintained by the water utility, or as in the case of new distribution system model developments, by the developer and/or their client.

Many utilities are moving their network records to GIS. A GIS provides the modeler with the opportunity to efficiently extract existing model input files, link to model node characteristics via a linked database, and also generate "up-to-date" models much faster.

For utilities without a GIS, the development of a detailed model of their water system is a first step in creating a GIS of their water system. Expanding the model data collection process to include the creation of a GIS slows the analysis process but ultimately results in substantial benefits to the water utility.

Paper Records. Paper record sources for distribution system data include the following:

- Water distribution maps
- Contour maps
- Aerial photography
- As-built construction drawings
- Operation and maintenance manuals
- Zone boundary maps
- Annual reports
- Planning documents containing information on the age of neighborhoods

All of the latest hydraulic modeling software packages have the ability to graphically present results of a distribution system analysis. These software packages also allow data entry using state-of-the-art digitizing techniques, often with aerial photos or other types of a land base that aid in positioning spatial data. In order to do this, however, the input files must include the geographic artesian coordinates (x-y) for each node. The coordinate system used by the utility's GIS should be used for the model to ensure spatial compatibility between the two systems. If the utility does not have a GIS system the coordinate system of the paper maps or other asset data should be used. A standard coordinate system, such as the US Geological Survey (USGS), State Plane, or Universal Transverse Mercator coordinate system, is recommended. If the model is an established coordinate system, then the model can be converted to a different coordinate system if the utility decides to change the systems. Using the same coordinate system as the rest of the utility's data simplifies model maintenance and enables the use of model results in conjunction with other utility information.

Electronic Records. The types of electronic data sources that are used to extract network data include the following:

- GIS
- SCADA
- DTMs
- Maintenance Management Systems (MMS)
- CADD drawings
- Database systems

A GIS generally contains more information than is required for building a hydraulic model, so the GIS data should be filtered before entering it into the model input file. A GIS requires that a unique pipe element be defined for each fitting, such as a pump or valve, in the water distribution system. The hydraulic model, in contrast, is only concerned with changes in the pipe network that result in significant changes in flow velocities and flow patterns.

The extra detail in the GIS requires development of data extraction routines that filter out this extra information and then combine the individual pipe segments into one hydraulic element. All segments in the hydraulic pipe element should be modeled to have the same diameter and roughness or friction factor. A new route is defined to identify a different pipe diameter, in-line fitting, etc. When extracting data from a GIS, the modeler should be aware of potential updates and future expectations of the model.

With the creation of a model from electronic data sources, there is an expectation that future model updates can be done quickly in response to infrastructure upgrades or system replacements. GIS extraction procedures and model input file generation software should be developed, allowing for timely updates.

SCADA represents the collection of equipment, communications, and software that allows the control and monitoring of system operation. SCADA systems are typically used to control the operation of the distribution system and to report on the behavior of equipment, as well as the distribution system itself. SCADA systems are designed so that pressures, flows, pump status, reservoir levels, chlorine residuals, valve position, and other information is collected at regular intervals, and stored in a database. The SCADA system therefore provides a wealth of information that is downloadable for use in the model or to compare against model results.

The Supervisory Control function can be manual or automatic. In manual mode, an operator at the control center operates valves, pumps, or other devices via a software interface. In automatic mode, SCADA software controls the devices based on a preset operating strategy. In practice, most SCADA systems have a combination of manual and automatic controls, depending on the field device and the system characteristics.

There are four main components of a SCADA system:

- Remote Terminal Units (RTU)—These components provide the interface to field devices. They perform some local functions, but their primary purpose is to facilitate communication with the centrally located SCADA system.

- Communications—The method of communication is a key component of the SCADA system. Reliable communication is necessary to ensure that the equipment operates correctly when a command is initiated and that accurate data is collected for historical trending analyses. Communications media include radio, metallic cable, fiber optic cable, cellular data packet, and satellite. Constraints on availability, cost, reliability, and location dictate the selection of the communication medium.

- Master Station—The Master Station is located at the central command location. This station scans all the remote RTUs to confirm operation, collects operating data, and issues operational commands.

- Human Machine Interface (HMI)—The HMI is the software that presents the information to the operator in a textual and graphical manner. Commands to the field devices are sent via this interface. The look of the interface is typically customized for each utility.

DTM is a three-dimensional representation of the ground surface. DTMs are used to assign elevations to nodes in a model. DTMs are usually used with CAD or GIS packages.

MMS are used by many utilities to track repairs, improvements, or other changes to a water distribution system. An MMS is often a source of valuable information regarding the water distribution network and related facilities and is, therefore, a source of information to create or update the model.

CADD systems were used to store drawings or maps of a water distribution system. Although more and more utilities use GIS to track utility assets, the CADD system is still a source of valuable asset information.

Utilities will often store information useful for model development in relational databases. Customer billing systems, for example, usually contain information about the water use of each customer. Some utilities store asset information, repair data, or other information in a database that may be useful to the modeler.

Physical Inspections. The physical characteristics of a facility provide information to the modeler on how the system operates. The modeler should conduct a field inspection to understand how the system functions, taking note of how the system is laid out as well as any operating peculiarities. This inspection will familiarize the modeler with the following:

- Number, type, and condition of the water supply facilities
- Types and age of pipes and overall condition of the piping network
- How system components perform over the range of demands
- How the components interface with each other
- Limitations and deficiencies of the system
- Recent work completed to improve system performance and efficiency
- High-volume water use customers that reveal critical flow information significantly affecting model performance
- Valve location and position (open or closed)

A site visit to water sources, pump stations, reservoirs, pressure-reducing valve (PRV) stations, and monitoring points is very useful to understand how the distribution system operates and helps ensure that site drawings actually represent current infrastructure.

The observed physical characteristics should be compared against the network and operations data to ensure that the pipes and connections are accurately represented. The location of all pressure and flow monitors should be confirmed. Elevations of pressure sensors should be compared to elevations being modeled. Control valves should be checked to determine whether or not they are at the correct set points. They should also be checked to assure that they are fully operational and not stuck in one position. Noises indicating pump cavitation should be noted, as a pump may be operating considerably off its original design point, either because of impeller problems, incorrect set points, or obstructions in the pipes affecting the operating point of the pump.

Pipes being replaced should be examined to determine the interior condition, assessing how roughness of the pipe has changed in time. This is then compared with the results of friction factor-testing programs (see chapter 3) to determine whether or not the values measured are consistent with observed conditions. A c-value test is conducted before and after replacement if the utility can afford to do so.

Large transmission main valve chambers may have small air release valves located on either the upstream or downstream end of the valve. If possible, a second air release port is added on the other end of the valve, allowing a quick check of pressure drop across the valve to determine whether the valve is fully opened.

Degree of Detail

The degree of detail required in the model is greatly influenced by the intended purpose of the model. A model used for long-range planning, transmission main sizing, and reservoir siting only requires modeling major transmission mains in the network, only rough estimates of demand distribution across the distribution system, and the use of theoretical peaking factors for estimating water demands. Depending on how the pump stations and reservoirs are operated, the level of detail collected at these locations will vary from simply knowing the set delivery pressure at each location to detailed pump and reservoir information that accounts for operation variability.

On the other hand, a model used for leakage control studies, energy-use management, or water quality modeling requires considerably more detail, perhaps including data for each and every pipe in the distribution system. It is necessary to know where the demands occur and what the associated peaking factors are throughout the day.

The amount of time and financial resources budgeted also influence the level of detail incorporated in the model. Where detail is lacking, it is critical that the modeler be aware of the limitations of the model—these limitations should be clearly stated in any report documenting analysis results. The modeler may be required to conduct a cost–benefit analysis to obtain managerial support for extending extra effort in developing a more detailed model.

Degree of Accuracy

The degree of accuracy required of the model input file parameters and of the system model depends to a large extent on the distribution system's sensitivity to change. For example, if relatively small changes in demand result in large changes in system pressures, characterization of pipeline hydraulics, allocation of demands, and representation of actual pump characteristics are certainly more critical than for a distribution system that is less sensitive.

After the model is built, it is recommended that the modeler conduct a sensitivity analysis on key model parameters to determine which factors strongly influence pressures and flows.

Skeletonization

Skeletonization is the process of stripping the network of pipes not considered essential to the analysis. Skeletonizing a system reduces the complexity of the model while still maintaining sufficient accuracy in simulating network hydraulics and reducing model preparation and analysis computation time.

The degree of model skeletonization depends on the purpose of the model and the size of the overall system. A transmission main study may only require the largest pipes in the network. Many master planning studies may only require pipes with diameters larger than 12-, 16-, or 24-in mains. Some small distribution models can work effectively with all the pipes in the distribution system. Model studies that focus on a particular portion of the distribution network may require full detail in the study area, but other portions of the model could be skeletonized or eliminated entirely. Water quality analysis is usually done with models that are not skeletonized, in order to most accurately represent water velocities. The scenario management capabilities now available with many software packages allow the modeler to store a complete set of pipes in the database, then extract only those pipes relevant to a particular study.

Summary of Model Formulation Steps

Model formulation is summarized in Figure 2-1. Subsequent sections in this chapter explain these steps in more detail and describe some of the methods used to collect and analyze data.

```
┌─────────────────────────────────────────────────────────┐
│  Determine Purpose of the Model and Simulation Scenarios │
└─────────────────────────────────────────────────────────┘
                            │
                            ▼
┌─────────────────────────────────────────────────────────┐
│   Establish Inventory of Existing Facilities and Data Sources │
└─────────────────────────────────────────────────────────┘
                            │
                            ▼
┌─────────────────────────────────────────────────────────┐
│      Review Operating Records and Interview Operations Staff │
└─────────────────────────────────────────────────────────┘
                            │
                            ▼
┌─────────────────────────────────────────────────────────┐
│   Review Consumption Records and Develop System Demand   │
│      Curves and Determine Critical Demand Conditions     │
└─────────────────────────────────────────────────────────┘
                            │
                            ▼
┌─────────────────────────────────────────────────────────┐
│  Compile Collected Data Into a Working Hydraulic Model.  │
│  Validate Model by Checking Model Results for Reasonableness. │
│   Calibrate Model Using Field Data to Demonstrate That It │
│       Represents the Actual Distributiion System.        │
└─────────────────────────────────────────────────────────┘
```

Figure 2-1   Model formulation steps

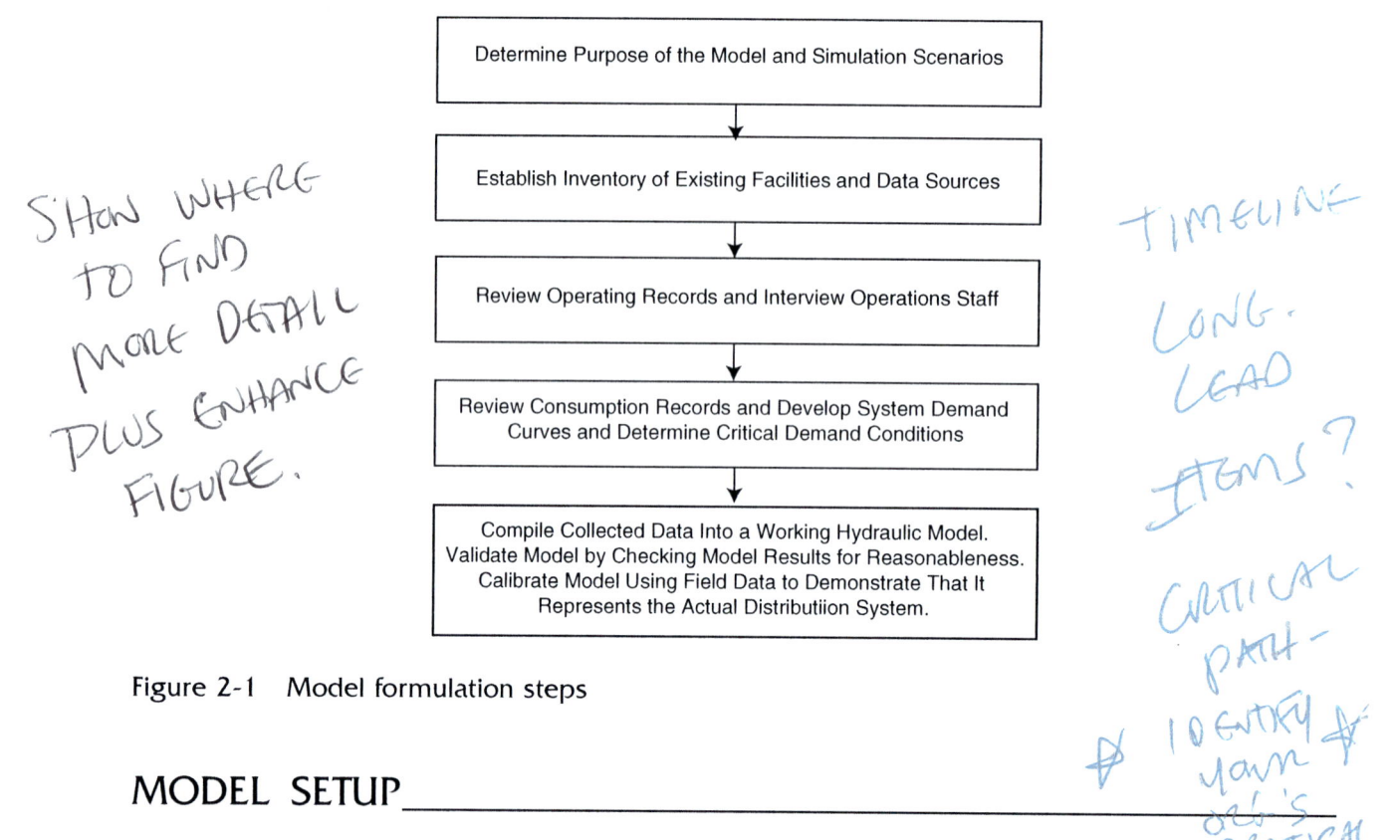

MODEL SETUP

Node Data

Nodes are points in a model that represent the end points of pipes, valves, pumps, as well as tanks and reservoirs. Nodes are where elevation information is stored, and where supplies and demands enter or leave the network. Topology within the network is usually defined by creating a logical link between the node and the facilities (called links or pipes) that start or end at a node. Nodes should be created at tees, crosses, plugs, reducers, and changes in pipe roughness or diameter, significant demand points, and at elevation extremes. Models report pressure, hydraulic grade line, and flows at node locations.

A node is where a water supply or a demand is represented in the model. Supply nodes represent water flows into the distribution system and model the treatment plant, groundwater well, or reservoir location. Supply nodes have constant or time-varying pressure or flow rate into the network. Demand nodes are located throughout the water network and are locations where customer demands are modeled. Depending on the software package, the demands are either assigned as a lump consumption representing all customer types or as a series of consumption values representing the different customer types, such as residential, industrial, and commercial classifications.

Nodes also have geographic coordinates that help establish the physical layout and representation of the network. Nodes have a data field where elevations (cartesian z-coordinate) of the water network are stored.

Elevation Data

Elevations used in the hydraulic analysis are separated into two categories: (1) control elevations; and, (2) ground elevations.

Control elevations are defined as measured system elevations at locations critical to the model, such as at pump or PRV stations. Pressure measurements at these locations are used in the model calibration process to determine whether or not the model is performing as expected. At these locations, the elevation assigned to the node should be the elevation of the pressure gauge and should be as accurate as possible. Control elevations are often determined by surveying.

Ground elevations are typically used for the remaining model nodes. These elevations are not required for calculating hydraulic grade line but are necessary for determining available delivery pressures in the water system. Most fire flow standards are based on available pressure at the hydrant (nearly at ground elevation) rather than the pressure in the main. This is the reason that ground elevations are recommended rather than pipe elevations. Ground elevations should be accurate within +/− 1.5 ft (0.5 m).

Hydraulic models that have relatively long distances between node locations and have substantial changes in topography may not have nodes that coincide with the high and low pressure points in the distribution system. In these cases, additional nodes are added to the network near these high and low locations in order to represent the extreme pressures. When analyzing a distribution system, elevations of surrounding distribution networks are also checked to determine if the elevations used by the adjoining model are consistent. Surrounding systems may provide boundary conditions for the model. A simple static pressure check is adequate to identify potential problem areas.

Use of high-quality global positioning systems calibrated to known benchmarks is a good method for obtaining accurate elevation data. As-built drawings and sewer system manhole elevations are also used for elevations. Elevations from DTMs are of widely varying quality and must be used with caution. Even though the elevations may be printed out to many decimal places, they are frequently extracted from sources that are only accurate to 20 ft (4 m) and hence are only as good as their source. When interpolating elevations, review the maps and identify any geographic break lines that limit the use of interpolation techniques.

FACILITIES DATA

Facilities data are the core component of a hydraulic model. Facilities data consist of all information that is considered reasonably constant in the analysis. This includes data on nodes, ground elevations, pipe lengths, pipe roughness, pump curves, valve characteristics, and reservoir characteristics. The pipe material and age are used to estimate pipe roughness.

All software packages used in hydraulic network analysis require the same basic facilities data. These model data are extremely valuable because of the value that the data provides the utility in terms of cost savings as a result of improved decisions and because of the cost of collecting and maintaining the data. This investment in data should be protected by ensuring that the data are easily taken directly or indirectly from one modeling package to a different modeling package if the utilities' modeling software requirements change. Most software vendors use standard, commercially available databases and file formats that enable easy transfer of data.

Pipeline Data

The pipeline data entered into a hydraulic model consists of diameter, pipe length, and an internal roughness factor. Pipe roughness factors are estimated from the diameter, material, and pipe age.

Pipe Materials

Over the years, new pipe materials were introduced and manufacturing methods were largely improved. As a result, many mature water systems are comprised of a wide variety of pipe materials with varying hydraulic properties.

Some of the pipe materials found in a water distribution system include:

- wood stave
- unlined cast iron
- lined cast iron
- ductile iron (cement lined)
- steel
- concrete
- asbestos cement (transite)
- polyvinylchloride (PVC)
- polyethylene (PE)

The following tables report roughness factors for these different pipe materials. It is important, however, to conduct local roughness factor testing to determine the extent of hydraulic capacity deterioration of the pipes, as hydraulic capacity is influenced by local water quality conditions.

Material Properties

Manufacturers of the different pipe materials typically provide roughness factor information for new, relatively clean pipes. In general, pipes are initially very smooth inside and perform hydraulically as described in the literature for a few years. Over time, however, transformations may occur within the pipes through a number of mechanisms. These mechanisms include corrosion, encrustation, and biofilm development. Corrosion and encrustation rates are dependent on the local water chemistry. These processes effectively reduce the interior diameter of the pipe, which is subsequently reflected in the model by adjusting the roughness coefficient or friction factor.

Length

The length entered into the hydraulic model is the length of pipe between node locations, including fittings. This length is obtained from maps or through digital extraction. In areas with substantial changes in ground elevation, care should be taken to enter the true total length of pipe and not just the horizontal distance between nodes.

Diameter

The true inside diameter of pipes is used whenever this information is available. Actual internal diameters provide better accuracy, especially for large transmission mains and for water quality modeling. The modeler should be aware of the difference between nominal and actual diameter when conducting flow tests or when comparing the economics of two different pipe materials.

For larger pipe sizes, the difference between nominal and actual diameter is significant as increased wall thicknesses are required in larger diameter pipes to maintain sufficient pipe rigidity. For transmission main modeling, the actual internal diameter is recommended over the nominal diameter. When the actual internal diameter is not known, an appropriate friction or roughness factor is determined if the transmission main is monitored during calibration.

Roughness

Pipe roughness is estimated by knowing diameter, material, and age (year of construction). A testing program is required to determine the appropriate correlation of these variables to roughness factors in the local conditions.

Many empirical formulas were developed over the years to characterize flow of water through pipes. Several of the older formulas were based on tests of pipes manufactured in the early 1900s. These earlier pipes were generally manufactured under less stringent quality controls than presently and were connected with many fittings that caused significant disturbance in the flow path. Two formulas, whose derivations were based on smooth pipe, are commonly used in water supply to calculate head loss or volumetric flow rate—they are the Hazen-Williams formula and the Darcy-Weisbach formula. Both formulas are available for use within hydraulic modeling software packages.

The Hazen-Williams formula is used mainly in the US, whereas the Darcy-Weisbach formula is used predominantly in the rest of the world. The main difference between the two formulas is that the Hazen-Williams formula assumes that the roughness factor is constant regardless of flow rate, whereas the Darcy-Weisbach formula adjusts the roughness factor based on the Reynolds number (a dimensionless numerical quantity used to characterize the type of flow in a hydraulic structure, where resistance to motion depends on the viscosity of the liquid in conjunction with the resisting force of inertia). Both equations are based on mean flow characteristics (steady velocity profile) and ignore turbulent (unsteady velocity profile) phenomena in pipes (Filion and Varney 2002).

Most software packages allow the use of both formulas in the same analysis.

The following two tables, Tables 2-2 and 2-3, list standard roughness factors used in initial model development.

Table 2-2 Hazen-Williams C-factors (American Society of Civil Engineers, "Pipeline Design for Water and Wastewater," 1992)

Pipe Material	C-factor (Hazen-Williams Coefficient)
PVC and PE	140 to 150
Ductile Iron, Steel, AC	135 to 140
Cement–Mortar Lined—good workmanship 24 in. and larger	120 to 130
Cement–Mortar Lined—smaller than 24 in.	115 to 125
Unlined Cast Iron	80 to 120

Table 2-3 Darcy-Weisbach friction factors for pipe roughness in mm ("Pipeline Design for Water Engineers," D. Stephenson 1976)

Finish	Smooth	Average	Rough
Glass, Drawn Metals	0	0.003	0.006
Steel, PVC, AC	0.015	0.03	0.06
Coated Steel	0.03	0.06	0.15
Galvanized, Clay	0.06	0.15	0.30
Cast Iron or Cement Lined	0.15	0.30	0.60
Spun Concrete or Wood Stave	0.30	0.60	1.5

Hazen-Williams Formula

$$V = 1.318Cr^{0.63}s^{0.54} \tag{2-1}$$

Equivalent metric equation:

$$V = 0.849Cr^{0.63}s^{0.54} \tag{2-1M}$$

The head loss h_f may be calculated from

$$h_f = \frac{4.72Q^{1.852}L}{C^{1.852}D^{4.87}} \tag{2-2}$$

Equivalent metric equation:

$$h_f = \frac{10.65Q^{1.852}L}{C^{1.852}D^{4.87}} \tag{2-2M}$$

Where:

- V = mean velocity, ft/sec (m/sec)
- C = Hazen-Williams coefficient
- r = hydraulic radius of pipe, ft (m)
- h_f = head loss [ft (m)] in pipe length L, ft (m)
- Q = discharge, ft^3/sec (m^3/sec)
- L = length of pipe, ft (m)
- D = nominal diameter of pipe, in. (mm)

Tests have shown that the value of the Hazen-Williams roughness coefficient C is dependent not only on the surface roughness of the pipe interior but also on the diameter of the pipe. Flow measurements indicate that for pipe with smooth interior linings in good condition, the average value of $C = 140 + 0.17d$.

However, in consideration of long-term lining deterioration, slime buildup, etc., a higher value is recommended: that is $C = 130 + 0.16d$.

Darcy-Weisbach Formula

$$H_L = f\left(\frac{L}{D}\right)\left(\frac{V^2}{2g}\right) \tag{1-3}$$

Equivalent metric equation:

$$H_L = 0.3048f\left(\frac{L}{D}\right)\left(\frac{V^2}{2g}\right) \tag{1-3M}$$

Where:

- H_L = head loss [ft (m)] in pipe length L, ft (m)
- f = Darcey-Weisbach friction factor
- L = length of pipe, ft (m)
- V = mean velocity, ft/sec (m/sec)
- D = nominal diameter of pipe, in. (mm)
- g = acceleration of gravity, 32.2 ft/sec^2 (9.81 m/sec^2)

PREPARING THE MODEL 25

The friction factor, f, is determined from the Moody Diagram, knowing the pipe relative roughness (ϵ/D) and Reynolds number (R), as described in AWWA Manual M11 *Steel Pipe—A Guide for Design and Installation*. Most modeling software packages require that pipe roughness be entered into the model, and then f is calculated automatically.

Pumps

Pumps lift water to higher elevations and provide adequate pressure, as required by the network layout and ground elevations. Most pumps used in distribution systems are centrifugal force pumps with either a constant or variable speed drive—some utilities have pumps with both types of drives (see Figure 2-2).

Characteristic Curves

A characteristic curve is the unique representation of discharge pressure versus flow rate for a particular pump. The size of the pump and the shape of the impeller determine the characteristic curve of a pump. Variable speed pumps have a family of curves, where each curve characterizes the pump at different operating speeds. In most cases, pump characteristic curves are available from the pump manufacturer or supplier. Existing pumps should be tested periodically to determine their characteristic curves, as impeller wear over time affects pump performance.

When designing a water distribution system and selecting pumps, the range of system operating conditions should be considered. A very flat pump curve results in large changes in flow rate for small changes in pressure. In contrast, a very steep curve produces relatively small changes in flow rates over a relatively wide range of pressures. Most modeling software packages allow the modeler to include the pump curve data directly in the model.

Valves

The valves used in a water distribution system are broadly classified into three main categories: (1) manual isolation valves, (2) check valves, and (3) automatic control valves.

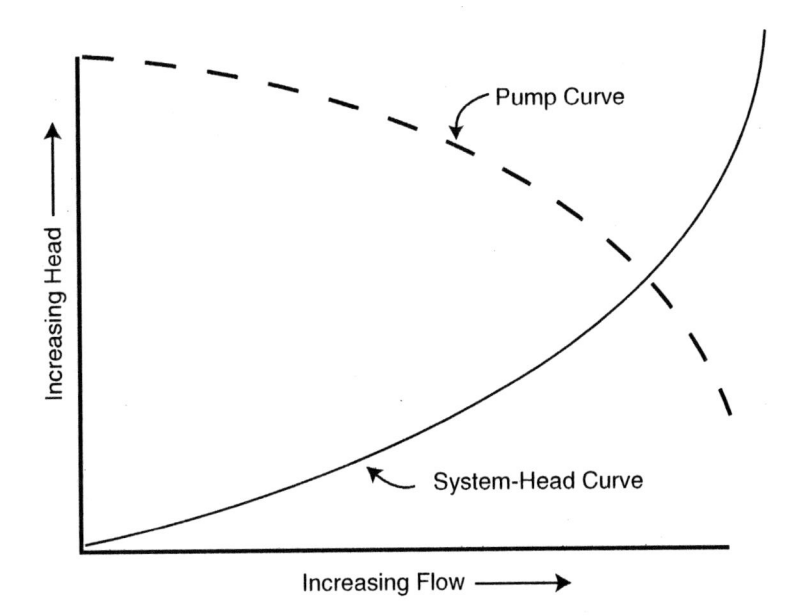

Figure 2-2 Pump curve

Manual Isolation Valves. Manual isolation valves are typically either gate valves or butterfly valves. They are located throughout the distribution system and within pumping facilities. In most cases, these valves are left either in the fully open or fully closed position and only operated during unusual events.

Manual valves are generally not modeled in the hydraulic analysis to decrease computation time. However, care should be taken, especially in pumping facilities where flow velocities are generally greater than 2.0 m/s (6 ft/s)—additional head losses through the valves are considered either by representing the valve as an equivalent length of pipe or, if the software allows, by inclusion of a valve element with an associated head loss coefficient based on percent open.

Check Valves. Check valves permit flow through a pipe in one direction only. They are used occasionally in distribution systems to prevent flow from moving in the wrong direction across a zone or other type of boundary. Check valves are located on the discharge side of many pumps, but these valves do not need to be included in the model because most modeling software does not allow reverse flow through pump links. However, check valves in distribution models should be modeled. Depending on the modeling software, the check valve should be modeled as a valve link or a pipe link should be defined to flow in only one direction.

Automatic Control Valves. There are five types of automatic valves typically included in water distribution system models: (1) pressure-reducing valves, (2) pressure-sustaining valves, (3) altitude valves, (4) flow control valves, and (5) vacuum breaking valves.

These valves have a sensing component, usually a solenoid valve, that opens and closes the main valve based on pressures or water levels in the distribution system.

PRVs detect pressure on the downstream end of the valve and will throttle if necessary to ensure that the downstream pressure does not exceed a set value. They are used in the distribution system where there is a substantial drop in topographical elevation—without the PRV, excessive pressures develop in the customer residences. When multiple PRVs service a single area, potential exists for the valves to work against each other, creating pressure waves as they try to reduce the operating pressures. Setting the pressure points to a slightly different grade line, making one of the valves predominate, sometimes alleviates this problem.

Pressure-sustaining valves measure the pressure on the upstream end of the valve and throttle to maintain a setback pressure in the system. They are often used on long transmission mains that terminate at a large reservoir. The reservoir feed can then be regulated to ensure that customers serviced along the main have adequate supply pressure.

Altitude valves are used in conjunction with reservoirs to automate the filling cycle of the reservoir. A level sensor opens and closes the altitude valves based on two (high and low) level set points. Altitude valves are either fully open or fully closed, so they can impose a large demand on a water distribution network when they open—for this reason, they are often used in conjunction with pressure-sustaining valves that throttle flow.

Vacuum-breaker valves are used to prevent backsiphonage when a partial vacuum pulls nonpotable liquids back into the supply line. The valve consists of a check valve operated by water flow and a vent to the atmosphere. When flow is forward, the valve lifts and shuts off the air vent. When flow stops or reverses, the valve drops to close the water supply entry and open an air vent.

Flow control valves are pressure-regulating valves installed to regulate the flow of water. Throttling the flow of water requires special valve designs that are durable over a long period of time. Butterfly and needle valves are commonly used to control flow in water systems in different applications.

Storage Tanks and Reservoirs

Storage tanks include both elevated and ground level storage facilities that supply water via gravity or via a pump station. Water levels in tanks fluctuate with system demand. Tank capacities, tank geometry, operating ranges, and control valves are important model inputs. For steady state analysis, much of this data is not included in the model but is required for use in interpreting and evaluating model results. In extended-time-period simulations, tank data is modeled so that water level fluctuations can be calculated and the altitude valves modeled correctly.

The capacity of a tank is represented as either the total volume stored or as the available volume stored. In many cases, the available volume is less than the total volume. System pressure requirements and fire and emergency storage reserve volume requirements may decrease the volume of water available during peak use periods.

Tank geometry, overflow elevation, freeboard constraints, and dead storage volumes, if applicable, should be determined for each storage facility. Either water level elevation indicating an empty tank or the minimum level limit for available storage should also be determined. Pressure switches, activated on tank water levels, possess control information for activating pumps.

Based on the tank dimensions, flow rate into or out of the reservoir is related to an increase or decrease in the water level, respectively. This relationship is important in determining whether a storage reservoir has sufficient capacity to meet specific demands (fire flows and peak demands) for a desired time period. For extended time period simulations, this information must be included in the input file so the system model can continuously adjust the tank water levels. Most software programs require specification of the operating range, as well as geometric information, such as volume versus water level for each tank. For steady state analyses, the modeler can calculate the change in storage volume using a volume per depth value for each reservoir. Reservoirs are represented as a large body of water where the water level does not change as a function of flow into or out of the reservoir.

DEMAND DATA

Demand data are added to the model after the facilities information is in place. There are three major steps that must take place in order to have properly allocated demands in the model. These steps are (1) determine demands, (2) allocate demands in the model, and (3) adjust demands to develop factors to convert from average day demands to the demands at the conditions that are to be modeled. There are multiple methods of obtaining demands and allocating these demands to the model. Several of these methods are presented in this section. The modeler will need to evaluate the modeling requirements, budget, and available data in order to select the most appropriate demand calculation and allocation method.

Most hydraulic analysis programs assign the customer demand to the nearest node. Advances in the software and in GIS technologies allow assignment of demands directly to the applicable pipe, at an appropriate distance along the pipe.

Allowing the steady state model to be adaptable to extended-time-period modeling, it is recommended that the input demands at each node be the average daily demands. Peaking factors are applied on each of the customer types to develop the maximum day and maximum hour demands.

Determine Demand

The first step in determining demand is to review unique characteristics of the water system. The information collected should include a listing of different customer types in the community (customer types consist of all customers having similar demand patterns). For example, a residential group might include all accounts expected to have minimal demand in the late evening and higher demands in the morning and early evening. Irrigation practices for the community should also be considered. Special industries should be considered, as they likely have unique consumption patterns that require special analytical treatment in the model.

The three customer types most often seen in hydraulic analysis are the residential, industrial, and commercial customer classes. In some communities, additional customer classes, such as multifamily, government, and institutional may be required. Large facilities, such as military bases, prisons, universities, and hospitals, should be kept separate from the other customer classes, because each of these facilities will have its own consumption pattern related to the nature of its business. The modeler will need to select the customer classifications most appropriate for the water system being modeled and the data available to develop the demands.

Residential Demands. Residential demands consist of domestic consumption and irrigation consumption.

Domestic consumption consists of water used in flushing toilets, washing, cooking, and drinking. Design standards have resulted in more efficient water fixtures being installed in new homes, resulting in reduced average day consumption. The age of a neighborhood and the renovation rates in a community are important factors when assigning average day demands to a node.

Irrigation consumption consists of water used on lawns and gardens. This component of consumption is community specific and dependent on yard sizes and climate. A number of communities have water conservation programs that specifically target outside water use. These conservation programs often have a significant effect on the magnitude of irrigation demand.

Industrial Demands. The industrial customer category consists of all large water customers and is typically comprised of manufacturing plants. In some cases, water service to the site is from two remote meters, so the allocation of demands to the proper location on the distribution system is important. Each industrial site is expected to have a unique flow pattern based on hours of production and water use characteristics. Seasonal variation is usually not an issue, but variability according to the day of the week may be significant.

A survey of the large customers in a community is generally recommended to determine their water use patterns and demand characteristics.

Commercial Demands. Commercial use consists of demands by stores, restaurants, gas stations, offices, etc. Commercial establishments may have fairly consistent daily usage rates and predictable fluctuations over any given day. The hours of operation and the type of building influence the demand pattern. Depending on the proportion of commercial developments in the community, it may be necessary to separate the commercial category into two groups—for example, delineate offices and shopping centers that may have different hours of operation or laundromats and car wash businesses.

Historical commercial flow data should be used with caution in forecasting future demands. A number of changes in the design of air conditioners, toilets, and other water fixtures have resulted in significant water cost savings for businesses. Commercial customers are likely to install new water conservation devices if the savings return period is less than one year. Again, the modeler should be aware of renovation changes in the commercial sector.

Water Loss

The difference between the total system input volume and the authorized billed and unbilled water consumption is the distribution system loss. If water loss is significant, it should be included in the model as a "dummy" demand at the nearest anticipated node.

Water loss is always present in a water system. Water loss consists of two distinct components: real losses and apparent losses. Real loss is the physical loss of water from the distribution system, including leakage and tank overflows. Apparent losses are "paper" losses, including meter inaccuracy, billing error, and unauthorized use.

Water loss results from several factors. These factors include unidentified leaks in pipes, main breaks, fire hydrant flushing, reservoir drainage for maintenance, unauthorized use, unmetered services, inaccurate and nonfunctioning meters, and on-site water plant usage.

Water plant site usage consists of water used for filter backwashing, chemical mixing, rinsing tanks, and sanitation purposes. In some cases, this water is not metered and can represent as much as 5 percent of the total water production. Neglecting losses attributed to plant usage does not adversely affect the model, as long as the volume leaving the site is metered.

Water loss in the distribution system must be added to demands in the model so that the total water supplied will equal the total water demand. The water loss is usually divided equally to all nodes because specific or isolated causes are difficult to determine, unless district zone measurements are made throughout the distribution system. System-wide district zone measurements allow a more accurate allocation of water loss. To increase allocation accuracy, some water utilities have used leakage tests in sub-areas of the distribution system for prorating the water loss on other areas having similar characteristics, such as pipe material, soil type, and age of main.

It is important to note that most steady-state analyses are based on the extreme conditions of peak hour or maximum day, plus fire flow. Under these conditions, the influence of inaccurately distributed water loss is much less than that under the assumption of average day demand conditions. Such inaccuracy is generally less than the achievable accuracy of the model demand allocation.

The amount of water loss in a distribution system may vary widely from system to system. Values ranging from 4 to 70 percent of the total water produced have been reported. Also, the amount can vary from year to year within the same system. The higher values are generally associated with communities that are not fully metered, have soil conditions that prevent surfacing of water from main breaks, or communities that do not maintain an active meter testing/replacement program.

Systems operating at relatively high pressure generally experience greater water losses, presumably through higher leakage at pipe joints. The hydraulic model is used to identify areas having potentially higher system loss and to develop operational strategies for reducing pressures, thereby reducing system loss.

Sources of Demand Data

Customer Information Systems. A CIS typically tracks the following information about each customer:

- Account number
- Name
- Address (billing & service)
- Meter-read consumption volumes

- Average water usage
- Date period for each meter reading
- Meter information (make, size, etc.)
- Classifications (customer/demand classes)

Core pieces of data useful to a modeler are usually available in the billing system. These core pieces of information are as follows:

- Location (service address) for each customer
- Water usage for each customer (for demand analysis)
- Classification of each customer (for demand analysis)

Utilities metering customer consumption have an excellent source of data for allocating water demands to the model nodes.

Most billing systems contain information on monthly or bimonthly water consumption, water use category, meter route number, and customer address. Some billing systems contain additional information, such as number of residences for a multifamily account or business type for a commercial account. It is possible to extract this information into a database software package, allowing for the use of automatic techniques to assign demands to nodes.

The consumption data collected from the billing system is collected over a period of a normal demand year (12 months of data). In some special case studies, a particular season is extracted (i.e., an especially hot summer).

Once the average daily and monthly demands for each billing account have been determined, these average demands are aggregated by assigning them to the nearest model node not on a transmission main or in another zone.

Estimated Demands. If accurate or complete consumption records are not available from the billing system, other techniques are used to estimate and assign demands to the model nodes. The three most common demand estimation techniques used are based on zone production data, population counts, and land use characteristics.

Often, all three techniques are used to develop estimated demand allocation. These techniques are also used in developing models for analyzing future growth scenarios.

Zone Production Data. A water utility should have a record of the total daily production supplied to the distribution system. This production is allocated across the hydraulic model using population and land use estimates, however a more accurate allocation is achieved by initially subdividing the area into individual pressure zones.

To calculate demands by pressure zone, a record of the flows entering and leaving the pressure zone is required. Flowmeters and changes in reservoir levels are used to estimate the total volumes consumed in each pressure zone by calculating a mass balance for each hour of the day within the pressure zone. Care must be taken to identify any automatic control valves or check valves that could open or close, moving water across an adjacent pressure zone boundary.

Once the production is allocated to each pressure zone, consumption data from meter records is assigned to the model nodes and then subtracted from the total zone water usage. The remaining water volume is leakage and is allocated to the remaining nodes in the zone.

The simplest approach in assigning the remaining volume is to assume that the volumes used are dependent on the area of the land being serviced. To calculate the node demand, first determine the consumption per area (gallons/acre) by dividing the remaining volume by the remaining area to be serviced. Each node is then assigned consumption values by multiplying its area by the calculated consumption per area value.

Using additional information such as population counts and land use characteristics, further refines the process of allocating unmetered water loss within a zone.

Population Counts. Population counts are used to estimate consumption in residential areas. Generally, population information is obtained from government census data, and in some cases, the local transportation department will maintain traffic count data by neighborhood for use in designing collector and arterial road systems.

The modeler assigns a population value to each node and uses a per capita multiplier to convert the population number to a consumption value. Per capita water consumption can vary from community to community based on lawn irrigation needs and on the age of the area. There are many studies available that discuss how to estimate per capita consumption for an area. AWWA Manual M22, *Sizing Water Service Lines and Meters,* includes information on conducting field sampling programs to determine per capita consumption rates along with key factors affecting the rates.

Future growth is also evaluated by multiplying increased population densities or new service area populations by per capita consumption rate.

Land Use Characteristics. Land use characteristics are used to estimate residential, commercial, and industrial consumption. Land use is usually defined by the local zoning regulations, and in most cases, there are more land use classifications than required for water use classification. Consideration should be given to the community's plans for future land use.

Residential consumption is determined by creating an average population per land use category and then using the per capita consumption to determine demands. Water consumption for a particular land use category is also derived from meter readings of customers within that land use classification. Commercial and industrial consumptions are usually estimated based on land area and consumption per area. In the downtown core, the ratio is based on smaller units of square footage of floor space. In a commercial/industrial area, it is more likely that the consumption is based on larger units of acres served.

Some communities have developed a GIS to track land use zoning. These systems are used to calculate demands and assign these demands to nodes. Land use characteristics are also used to forecast water demands in the future, to evaluate master planning scenarios where historical data are not available to base demand calculations. Demands per unit land area of a particular land use classification are developed from existing customer meter data or from per capita water use estimates.

Demand Allocation Process

Demand allocation is the process of assigning water consumption values to the appropriate nodes in the model. The objective is to distribute demands across the model to best represent actual system withdrawal. Assigning demands to the model nodes often requires a good working knowledge of system usage. The hydraulic model needs realistic demands located at or near their actual locations on the network in order to predict the pressure-flow hydraulics of the network. This section discusses the different methods of linking customer records in the CIS to the hydraulic model.

Assigning customers can be a highly automated effort, a semiautomated effort, or a manual effort, depending on the quality of the service address information in the CIS, the land base used for the hydraulic model, and the availability and quality of commercial or governmental geo-spatial address information. The method of allocating demands depends to a great extent on what records are available to the modeler. In many cases, the final allocation is achieved through a combination of techniques and record sources. Demand allocation techniques utilize billing consumption records, meter routes, zone production data, population counts, geo-coded meter locations, and land use characteristics.

Many utilities now have customer meter locations in their GIS, so average demands associated with a meter are allocated to the nearest pipe or node using GIS spatial analysis tools. In some cases, polygons are drawn around each node in the GIS, and all meters whose coordinates fall within the polygon are assigned to that specific node. Land usage information in the GIS is used with customer billing data to develop demands that are allocated directly to nodes in the model using either a GIS or modeling software capabilities. GIS-based demand allocation techniques will be the most common methods of demand allocation because of the growing amount of geo-spatial data and number of GIS tools that are available.

Utilities that have a GIS in place to manage the tax parcels of the city can utilize the address and spatial location of each parcel for the hydraulic model. The hydraulic modeler could use this information to link the land parcel to the demand from the CIS. Once the spatial coordinates are determined for each customer, a spatial join is used to link each customer demand to the nearest pipe or node in the hydraulic model. Care should be taken to ensure that demands are allocated to the correct pressure zone near the zonal boundaries.

One traditional method used by water system modelers in assigning customer demands to nodes is via meter-reading routes. The utility may have a map showing the boundaries of each meter-reading route. From the map, the average demands per water use category are summed within each meter-reading route area. The location of each route is compared to the model network and percentages within each route area are assigned to each node. The water demand by water use category is determined for each node. Industrial and wholesale customers are considered special point loads and are assigned directly to the nearest node. If the meter route map is available electronically and each route is drawn as a GIS polygon entity, it is possible to develop software routines to automate the assignment of demands to nodes.

If an electronic meter route map is not available, the modeler should explore other options for assigning the billing records to model nodes automatically. It is now possible in most communities to purchase electronic street maps and postal code boundary maps used in GIS applications. These maps are generally created for local marketing analyses and may include other demographic information that could be used. It may require some probing to locate these maps, however, the time saved more than pays for the effort expended in locating them.

Electronic street maps have the capability to assign or map each customer's location based on the billing record address. The modeler ensures that the coordinate system used by the street map conforms with the coordinate system defining the model node locations. The model node locations and the individual customer locations are compared to determine the average demand per model node. Once again, the large industrial and wholesale customers should be treated separately to ensure that demands are properly allocated. If an electronic street map is not available, a postal code boundary map is used instead. These maps can be used following the same percentage assignment techniques used with the meter route maps. The modeler should be careful to ensure that the addresses used are street addresses, not billing addresses, in situations where the two addresses are not the same.

Adjusting Demands

Once the customer demands have been calculated and allocated to the model, the next step is to determine how demand varies according to location and according to time.

As stated previously, it is recommended that the base demands collected for the model input file be allocated based on either the average daily usage or maximum day usage. Variations in demand of these values are calculated by applying peaking factors based on location of the customer and on daily flow pattern. Depending on the scenario evaluated, it may be necessary to determine seasonal variations in demand. SCADA data is invaluable in obtaining both seasonal and daily factors to adjust average daily demands.

Geographic Variations. Just as the demand varies by the type of customer, it also varies depending on the neighborhood. The average size of the dwelling, the average number of water-use fixtures, and the size of lawns result in large variations in average residential demand between old and new neighborhoods.

Studies have shown that the per capita domestic consumption of new residences with water-efficient fixtures is less than the consumption of older homes with 15-year-old fixtures. This type of variation should be considered when developing a water distribution system model, especially when planning for system expansions and growth.

Water distribution system models are used to identify areas where the utility may want to concentrate an aggressive water conservation campaign to reduce the need for additional transmission infrastructure or to indicate that growth on the system will not require large capital expenditures.

Some utilities apply a different peaking factor pattern to their commercial customers if they are located in a commercial area or in the downtown core.

Local consumption patterns and characteristics should be reviewed by the modeler to determine if a local geographic peaking factor should be applied on the average demands.

Seasonal Variations. Seasonal variation is sometimes significant in distribution system models. Typically, seasonal variation is directly related to irrigation and air conditioning needs. A community with a large university or tourist season, should consider the institution's or seasonal changing needs when building a distribution system model.

Local climate variations determine the irrigation and air-conditioning requirements. The modeler should consider the return period of extended hot weather to determine what peaking factor should be applied. Irrigation and air-conditioning use are often targeted by water conservation efforts, as they significantly influence demand during extreme conditions and drive decisions on the required sizing and timing of water plant, reservoir, and transmission main expansions. The hydraulic model helps in assessing the effectiveness of a water conservation program.

Diurnal Variations. Diurnal demand allocation indicates demand variations throughout the day. This information must be obtained to establish peak demand factors. Diurnal demand characteristics vary based on customer type and geographic location. Typically, a demand curve is developed illustrating how the daily consumption varies over a 24-hr period, and this diurnal demand is compared to the average demand. The average daily demand is the total water usage over a 24-hr period, divided by 24 hr. Depending on the modeled scenario, the modeler selects the appropriate peaking factors for each customer type and location.

Field-testing and analysis of actual system operating data is required to determine the specific diurnal demand curves for the community. As a minimum, curves are developed for major geographic areas and pressure zones within the system. Curves are also developed for large industrial or wholesale customers.

To develop a diurnal demand curve for the overall system, water source flows from treatment plants and groundwater wells must be recorded, in addition to storage inflow and outflow. Ideally, this information is readily obtained from the utility's distribution SCADA system. A diurnal curve is developed by summing the inflows and outflows in a discrete area for each hour of the day to calculate the demand throughout the day. This same mass balance technique is used to develop curves for hydraulically isolated pressure zones.

Diurnal demand curves for specific customer types, such as low-density residential, high-density residential, and commercial is determined by isolating a homogeneous land-use area and monitoring the flow over a specified time period, typically 24 hr. Advances in flow monitoring (i.e., bidirectional direct-bury flowmeters) help identify areas for monitoring, and equipment can be left in place over many years without having to physically close valves and compromise firefighting capabilities of the system. Alternatively, strap-on sensors can be used in specific areas and locations to provide diurnal curve information for specific customers or groups of customers. These sensors convert meter dial movement to a digital signal. The diurnal demand curve for the service area under consideration is derived by summing all flows into and out of the area, measured by meters immediately surrounding the area (see Figure 2-3).

Diurnal patterns also vary significantly by season. The modeler should consider this factor when developing a sampling program to ensure that the data collected is suitable for the scenario modeled. For example, a winter sampling program is not useful for a hydraulic model assessing transmission requirements to meet peak hour summer demands.

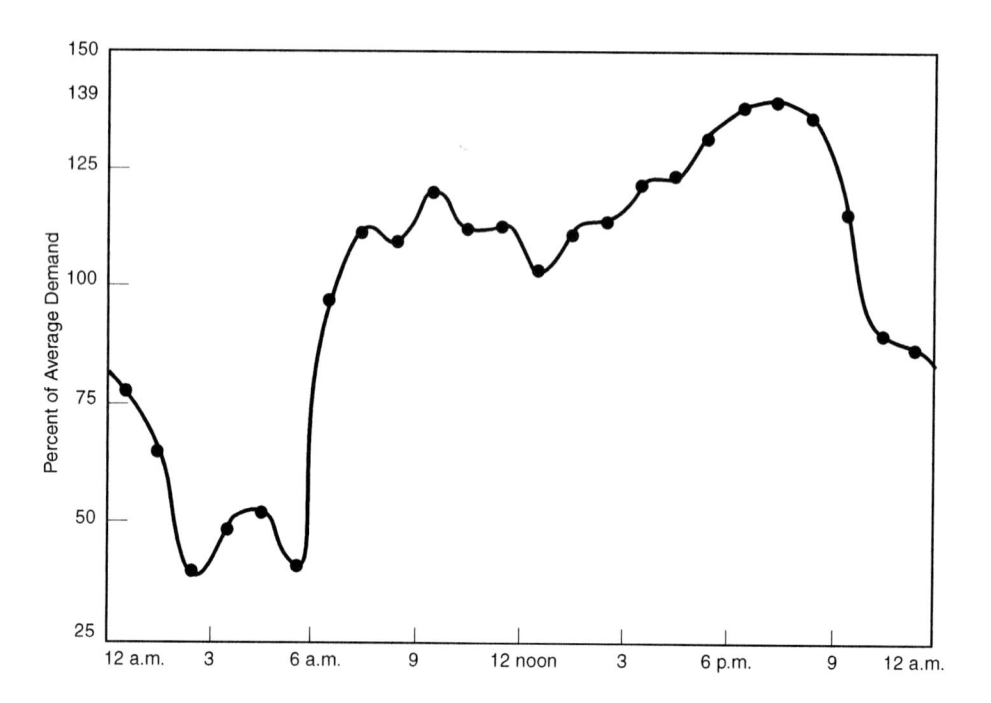

Figure 2-3 Diurnal curve

OPERATIONS DATA

In addition to network and demand data, the modeler must also collect information on the operation of the system being simulated. This data is referred to as operations data.

Operations data consists of all system parameters that change frequently during system operation and hence require careful consideration and modification when conducting hydraulic simulations. For steady-state analyses, applicable set points and operating conditions must be determined. Care must be taken when adjusting a steady-state model to a new operating condition, so that the new operating parameters are properly considered. In an extended-time-period model, the modeler inputs control logic data that defines the operating set points.

Operations data is generally classified as two main types: (1) stable (static) data and (2) dynamic data.

Stable data consist of items such as pump and valve control logic and set points for pumps, control valves, pressure-reducing valves and for altitude valves. Most software packages include some type of control scheme that replicates the behavior of a real control system. The model automatically changes valve position or pump status in response to changes in pressure, flow, or water level at specified points in the model.

Dynamic data consist of data that change based on an operator or supervisory control decision. The control logic for these data is often difficult to model in a system that has many different operating options (i.e., multiple choices of pumps to initiate at different locations to resolve a low-pressure problem). In these cases, the modeler must determine what appropriate combination of operations data is loaded into the model. Some modeling software packages load and modify this information through an additional input file.

Typical operations data required for distribution system models are listed in Table 2-4. The items are often interrelated, meaning model input changes may automatically result in recalculation of other model parameters. Operations data are used extensively in the model calibration process, where operations data are often used to define model conditions so that the model inputs and outputs match actual measured values.

There are four main sources of operations data: (1) operations staff, (2) written records, (3) charts, and (4) SCADA.

Table 2-4 Operation data required by facility/equipment type

Facility/Equipment	Operation Data Required
Reservoirs	• Water elevations (i.e., dead storage, overflow) • Inflow/outflow rates
Pumps	• Pump status (on/off) • Pump speed (variable speed only) • Flow rate and head
Pressure-Regulating Valves	• Upstream pressure • Downstream pressure • Valve position and setting
Flow Control Valves	• Valve position • Flow rate through valve • Downstream pressure
Node Pressures	• Any pressure data collected at pressure-monitoring stations in the distribution system
Pipe Flows	• Any in-line flow data measured in the distribution system (includes wholesale supply locations or zone boundary meters)

Operations Staff

The most important source of information in operating a water distribution system is the operations staff. The modeler should develop a good working relationship with this group in development of any hydraulic models. Operations staff can inform the modeler of how the actual system operates as opposed to how it was originally designed to operate. They can identify specific areas to study to determine why the actual operation varies from the design, such as finding closed valves, determining incorrect set points, and identifying distribution system deficiencies. During model calibration, operations staff input helps in interpreting the data and identifying necessary model adjustments.

Involvement of the operations staff in model development greatly enhances the acceptance of analysis results and may motivate more efficient operation of the water distribution system.

For these reasons, involving the operations staff in model development results in a more accurate and usable hydraulic model.

Written Records

Written records consist of all the reports and documents that explain how the distribution system is theoretically supposed to operate. The modeler should verify these reports against actual operating conditions, which may be different because of the presence of constraints and conditions in the distribution system that may not have been present or recognized when the reports were written.

Typical written reports that should be reviewed are

- Operations and maintenance manuals for each facility
- Design reports and record drawings
- Operator training manuals
- Emergency manuals (to identify how things are expected to operate in an unusual situation)
- Historic records of deficiencies in the system, i.e., low-pressure complaints, water quality complaints, etc.
- Maintenance records for all facilities, including pump stations, reservoirs, valves, pipes, etc.
- Valve status reports
- Hydrant flushing reports

Each utility will have a different combination of written records and in many cases the quality of the reports may not be consistent. As stated previously, operations staff offer the best assistance in determining the quality of the data collected from the written record.

Charts

Charts are historical records showing time relative data. Traditionally, these have been paper records and are generally in a circular or strip format. Digital data-logging equipment has reduced the need for paper charts, but software is still required to graphically represent and interpret the collected digital data. Care should be taken to ensure that charts and digital data are calibrated, so that the data is reliable.

In most cases, charts are used to record flows and pressures against time. The input of this data into the hydraulic model requires that the modeler manually compile it into an appropriate format.

One caution in using digital chart data to compare pressures and flows between two locations is that standard database and spreadsheet packages may not be suitable to handle the data. This is mainly because the data is time variant, and depending on how frequently the sampling is conducted, it is often difficult to compile and compare data from two unsynchronized locations. Software packages are available from data-logger equipment suppliers that help in analysis of the data. These packages are worth the cost to reduce the effort required when using standard databases and spreadsheets for analysis.

SCADA

SCADA systems are a powerful tool for collecting and maintaining historical information on the actual operation of a water system. This information includes flows, pressures, alarms, or tank levels and equipment information such, as the on/off status for pumps. The data is collected via the SCADA software and stored in database files for evaluation. The amount of data collected is determined by the polling frequency of the SCADA system. It is also possible to incorporate mathematical formulas to track combinations of data—typically done to determine total system demand based on a mass balance of the flows from the treatment plants and in and out of the reservoirs.

Some utilities do not make full use of the data acquisition capabilities of SCADA systems. This is unfortunate, as the data is of great value to the hydraulic modeler and the operator in identifying system bottlenecks and in improving operational strategies to reduce energy costs and enhance water quality.

Importance to Modeling. Historically, only the most critical operating points were monitored via SCADA. This is because of the increased cost of additional field devices and communication lines. These costs have since declined, therefore the modeler should review the data being collected and possibly arrange to have additional monitoring points installed at key locations.

SCADA provides a wealth of data on the actual operating conditions of the water distribution system. Differences between the model results and SCADA readings are used to recalibrate the model or identify locations where field crew checks are conducted to locate closed valves.

If sufficient data is collected, the SCADA easily generates data files for model calibration and allows the modeler to confirm the accuracy of the model on a regular basis. Usually, models are calibrated at the outset and then every few years thereafter. Some model software packages accept automatic SCADA updates and will generate log sheets identifying where the model and the system differ.

Developing common interfaces and linkages between the model and SCADA system improves the ability of the model to consider operational scenarios. This is accomplished by entering current operating conditions into the model and allowing the operator to perform "what if" analyses to test different scenarios. For example, different pump and valve combinations can be tested from the same operational starting point.

Data Requirements From SCADA. There are two types of SCADA data that can be used in modeling:

- Boundary condition data—these include information on which pumps are operating, flow and pressure control points, valve statuses, reservoir levels, and demand loading. This information is used in hydraulic calculations.

- Verification or reference data—these includes information collected by flow, pressure, and level monitors and are used to confirm that the model is sufficiently representing actual operation. These data are not used in hydraulic calculations.

It is important to note that SCADA data can include transient or surge data (transients are rapid changes in pressure and flow) that, if misinterpreted, can provide a misleading perspective of the average steady-state conditions of the network. Transient data is typically generated during pump startup and shutdown or during changes in valve status. The use of average data instead of instantaneous data values will help ensure that the data being placed into the model is a true representation of the pressures and flows in the system. A review of data values in adjacent time periods is also useful to identify anomalies in the data.

Data errors also occur during a communication failure. These should be flagged in the SCADA. When assembling data for the model, it may be necessary to assume default values if some measured values are not readily available.

Issues. There are three main issues that should be addressed when linking a hydraulic model and a SCADA system:

- Data translation—the equipment identifiers tracked in the SCADA system are usually not the same as the identifiers used in the model. A lookup table needs to be developed and maintained as changes are made to the model, field equipment, and SCADA system. This lookup table is used to map SCADA points to specific model locations.

- Data transfer—the data formats for the system model and for the SCADA are usually dissimilar. This requires the use of ASCII file transfers to move data between systems. The generation of the ASCII file format requires more advanced modeling and SCADA knowledge. The ideal transfer method is to have a programmed routine that the casual user can use to pull the data from SCADA. This routine should include data quality checks to flag the user when an incomplete data set has been extracted.

- Data conversion—the data collected in SCADA can be in different units than those in the model. Therefore, the data translation process should include a units conversion where necessary. Recoding clocks for different monitoring stations should be synchronized.

REFERENCES

Bonema, S.R., Elain, C., Mercier, M., and V. Tiburce. 1995. Linking SCADA with Network Simulation. In *Proceedings of the AWWA Annual Conference*. Denver, Colo.: AWWA.

Cesario, A.L. 1995. *Modeling, Analysis and Design of Water Distribution Systems*. AWWA, Denver, Colo.

Chase, D. and G.L. Jones. 1994. Linking Hydraulic Network Models and SCADA Systems: The ABCs of System Integration. In *Proceedings of the AWWA Computer Conference*. Denver, Colo.: AWWA.

Deagle, G. and S.P. Ancel. 2002. Development and Maintenance of Hydraulic Models. In *Proceedings of the Information Management and Technology Conference*. Denver, Colo.: AWWA.

Green, D., and C. Montgomery. 1998. SCADA Communication Experience at the Detroit Water and Sewage Department. In *Proceedings of the AWWA Annual Conference*. Denver, Colo.: AWWA.

Fortune, D., Campbell, P., and R. Cavor. 2000. Cost of Ownership of Hyrdaulic Models. In *Proceedings of the Infrastructure Conference*. Denver, Colo.: AWWA

Hutchison, W. 1991. Operational Control of Water Distribution Systems. In *Proceedings of the AWWA Seminar on Computers in the Water Industry*. Denver, Colo.: AWWA.

Japan Water Works Association, AWWA Research Foundation. 1994. *Instrumentation and Computer Integration of Water Utility Operations*. Denver, Colo.: AWWA

Mau, R.E., Boulos, P., Heath, E., Brennan, W.J. 1996. Advanced Network Modeling Applications: Dynamic Design and SCADA Interface, In *Proceedings of the AWWA Engineering and Construction Conference*. Denver, Colo.: AWWA

Rehnstrom, D.J., Butler, C.L. 2001. Maximizing the Use of Your Hydraulic Model. In *Proceedings of the Information Management and Technology Conference*. Denver, Colo.: AWWA

Schulte, A.M., Bonema, S.R. 1995. SCADA Data for Predictive, Training and On-Line Water Distribution System Simulation. *Proceedings International Water Supply Association, Specialty*. Denver, Colo.: AWWA

Wilson, L.L., Moshavegh, F., Bolze, M.A. 2001. Constructing, Maintaining, and Utilizing a Large Water Model. In *Proceedings of the Information Management and Technology Conference*. Denver, Colo.: AWWA

Wood, P. 1994. Application Integration for Improved Utility Operations. In *Proceedings of the AWWA Computer Conference*. Denver, Colo.: AWWA.

1989. AWWA Manual M11, *Steel Pipe—A Guide for Design and Installation*. AWWA, Denver, Colo.

This page intentionally blank.

AWWA MANUAL M32

Chapter 3

Hydraulic Tests and Measurements

INTRODUCTION

Hydraulic tests and measurements are important to modeling work because tests are used to obtain information for the model and to verify information from other sources. Hydraulic tests are also an integral part of model calibration. Many times, pieces of information are not available without taking measurements in the distribution system. Data such as pump curves, pipe friction factors, flow rates, customer demands, and reservoir levels are obtained from a field test when no other source of information is available. Also, there are times when information from other sources is incorrect, and the model does not match calibration data. Additional field testing is used to determine why the network does not behave as expected. In the process, closed valves, incorrect records data, severely tuberculated mains, and other problems are found and resolved, helping the utility improve operation of the distribution system. By using field testing to improve model calibration, confidence in the model increases, and the model becomes a trusted planning, design, and operational tool.

This chapter covers the following areas:

- Planning field tests
- Flow measurements
- Meter calibration
- C-factor tests
- Diurnal demands
- Hydraulic gradient tests
- HGL tests
- Fire flow tests

PLANNING FIELD TESTS

To be effective, field testing should be done with a well-thought-out plan. Field testing is expensive, and the test's potential for failure is high when there has been insufficient planning, when the parameters affecting the test are poorly understood, when insufficient data is collected, and when test equipment does not function properly. When planning a field test, first establish the objective, which may be to calibrate a model or a portion of a model, estimate C factors, check meter calibration, or to identify the cause of anomalies in the distribution system. It is often helpful to use a map of the network and draw out zone boundaries, transmission mains, and other facilities. This map can be annotated with the locations of fire hydrants to be tested, temporary meter locations, areas with unlined pipe, and known problem areas. Some tests require valves to be closed or other operational changes, which are also noted on the map. Areas where flow or pressure monitors are placed are marked. When possible, it is often helpful to use pressure or flow monitoring points that are on the SCADA system to reduce the requirements for field monitoring equipment. All the parameters that need to be monitored are identified. For example, reservoir levels, pump station flows, and pump station pressures provide valuable information during fire flow tests.

Flow monitoring is done at all the inlets and outlets of the network or pressure zone being tested so that the modeler has the information necessary to calculate a mass balance of the distribution area. Flow monitoring devices are placed at export locations, pump stations, PRV stations, and reservoirs. In some cases, flows are monitored into a subdivision in order to calculate the demands and diurnal demand pattern of a particular customer class.

After the requirements of the field test are established, an assessment of the equipment necessary to carry out the test should be made. Some utilities may have a few pressure data-loggers mounted on fire hydrants. Equipment can also be rented or purchased, depending on the need. The field testing equipment should be checked and calibrated when necessary. A one or two day preliminary test to ensure that the equipment works, and to ensure that a valid mass balance is obtained has often saved a full test from failure. Batteries should be fresh and digital memory storage must be adequate for the test. The time and time duration of the test should be determined. Tests lasting a week are common, to provide information over several days, in the event that operational problems on one or two of the days do not invalidate the test. Weekdays are usually the time when data is extracted for calibration, because demand patterns are often different on the weekend. Third parties, such as the local fire department, are notified so that equipment is not removed from fire hydrants prematurely. It is important to discuss the test with the control room operators so they will know to support the testing and help to make it a success.

Some tests are planned for time periods when the demands are high or at times when water velocities in transmission mains are high, so that there is sufficient head loss in the mains to get head losses greater than the accuracy of the instrumentation. Otherwise, the field test fails to yield meaningful results.

FLOW MEASUREMENTS

Flow measurements are an integral part of meter tests, C-factor tests, diurnal demand measurements, pump tests, and fire-flow tests. Flow-measuring equipment in water supply includes hydrant pitot gauges, pitot tubes, master meters, and portable magnetic meters.

Hydrant Pitot Gauges

Hydrant pitot gauges (Figure 3-1) measure the discharge indirectly by measuring the velocity head with a pressure gauge. The flow equation for a pitot gauge is

$$Q = 29.83 \times p^{0.5} \times d^2 \, C_H \qquad (3\text{-}1)$$

Where:

- Q = rate of flow in gpm
- p = gauge pressure of the flow stream in lb/in².
- d = diameter of hydrant discharge port in in.
- C_H = hydrant discharge coefficient

$Q gpm \times (6.3 \times 10^{-5}) = Q m^3/sec$

The magnitude of a hydrant coefficient depends on the physical transition from the hydrant barrel to the discharge port—typically 0.9 for round and smooth, 0.8 for sharp and square, and 0.7 for ports protruding into the hydrant barrel. Some hydrant pitot gauges are attached to diffusers, whose discharge coefficients may be obtained from the manufacturer or supplier.

Hydrant pitot gauges should be calibrated using a deadweight tester. The gauge resolution must be within 0.5 psi (3.45 kPa) to accurately measure pressure heads below 10 psig (68.95 kPa). A flowing hydrant cannot be used for a system pressure measurement because of friction loss in the hydrant valve and barrel. Therefore, pressure measurements should be taken from the two hydrants upstream from the flowing hydrant. Be aware that high discharge velocities from a hydrant damages lawns.

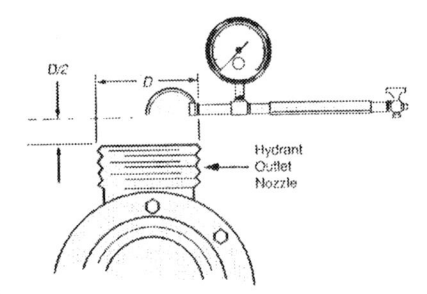

Figure 3-1 Hand-held pitot gauge

Pitot Tubes

Pitot tubes measure velocity heads within a pipe. Velocity heads are typically small for nominal flows in water mains. For a flow velocity of 0.5 fps (0.15 m/s), the equivalent velocity head is only a 0.05 in. (0.0013 m) column of water, which is difficult to measure. To get a good test, generally higher than normal flow should be induced. Theoretical velocities are adjusted with an instrument coefficient from laboratory calibrations.

Flow rate in a pressurized pipe is computed by multiplying the cross-sectional area of the pipe by the average velocity of the flowing water. The inside diameter is used to compute the cross-sectional area, whose value is adjusted for the obstruction of the pitot tube, as well as any corrosion or tuberculation inside the pipe. The average velocity may be determined from a velocity profile.

Figure 3-2 illustrates a method for determining velocity profiles in which a pipe is divided into 5 rings of equal area. Figure 3-3 shows typical profiles at two different flow rates in the same pipe. The average velocity is the mathematical average of 10 measurements (5 measurements on each side of the pipe center). At a particular gauging point, the ratio of the average velocity to the center velocity is called the *velocity factor* (VF), which is nearly constant across the normal range of flow velocities observed in water distribution systems.

Insertion points for pitot tubes are usually at 1-in. (25.4-mm) corporations or at air release valves. The best locations for stable velocity profiles are along pipes having at least 10, and preferably 20, diameters of straight pipe upstream of the gauging point.

Master Meters (Propeller Meters)

Master meters consist of a propeller mounted axially in the line of flow within the pipe. As water passes through the meter, the flow induces a rotation of the meter, which is calibrated to measure the flow-rate. These meters are usually installed permanently in a pipeline.

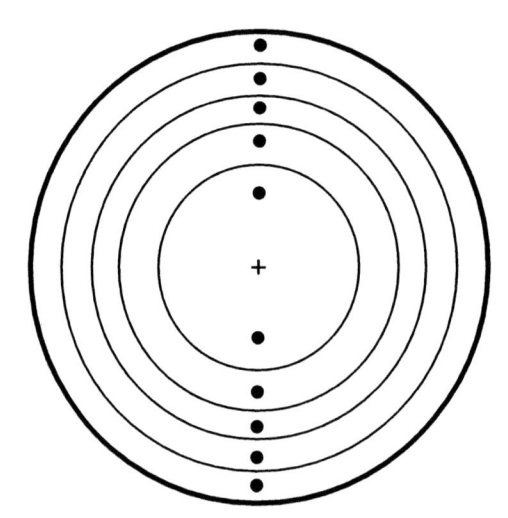

Figure 3-2 Traverse positions within a pipe

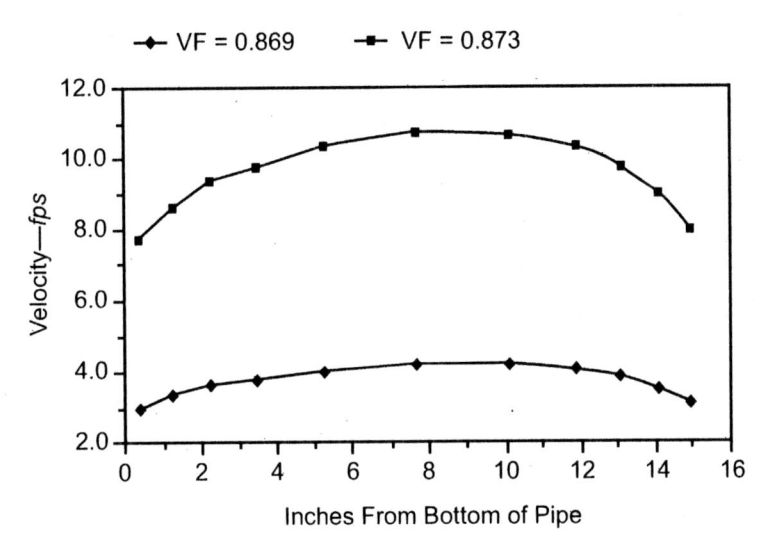

NOTE: fps × 0.305 = m/sec
In. × 25.4 = mm

Figure 3-3 Typical velocity profiles at two different flow rates

Magnetic Meters

Magnetic meters consist of a coil of wire around the pipe through which a current is induced as ions in the water pass through the pipe. These meters are calibrated so that the induced electrical current correlates to the flow rate in the pipeline. Magnetic meters are usually installed permanently, although temporary installations are also possible.

METER TESTS

A meter test compares master meter registration to flow rates measured by another instrument, such as a test meter or an inserted probe. Meter tests are used in verifying production records, in measurements such as pump tests, and in calibrating computer models.

Meter tests compare meter registration to the measured flow in a pipe, rather than simply checking electronic calibration of a transducer. Many meters include a primary device, such as a venturi tube, and a secondary element, such as a transducer, which converts the physical output to an electronic signal. A malfunctioning primary device causes meter error even if the secondary element is calibrated correctly. For example, a venturi meter may provide inaccurate readings because of leakage between the venturi tube and transducer, even though the transducer is functioning correctly.

Meter tests are conducted under normal operating conditions. At high flow rates, a meter may provide readings that are sufficiently accurate, but the same meter may provide inaccurate readings at lower flows that are more typical of normal operating conditions. Conditions in a shop test may not duplicate influences caused by upstream piping, turbulence, or other field conditions.

C-FACTOR TESTS

C-factors are friction coefficients used in the Hazen-Williams formula (see chapter 2). Hydraulic models use the Hazen-Williams formula or other head loss equations, relating flow to head loss in each pipe element within the model. The Hazen-Williams formula, although empirical, as are Moody diagrams and other equations, provides sufficient accuracy within the normal range of velocities and temperatures in a water distribution system.

Measuring C-factors for each and every pipe is usually not practical; therefore, assumptions are made based on a sample of C-factor measurements. The sampling includes all combinations of pipe sizes and materials, and also old and new pipes. Larger pipes tend to have higher coefficients. If there are unlined cast iron pipes in the system, they should definitely be included in the sample, as C-factors vary widely from less than 25 to over 100, depending on the pipe age and water quality. A map from the 1940s helps to identify unlined pipes because the use of unlined cast iron pipes ceased in the 1950s. It is important to take C-factor measurements near sources where treatment processes have coated the inside of pipes.

Hydraulic models usually use C-factors that account for losses in bends, valves, and other fittings. Losses at these fittings are usually small at normal velocities in water distribution systems.

The procedure for conducting field tests involves measuring flow, head loss, diameter and length, and solving the Hazen-Williams formula for the C-factor coefficient. The formula for computing the C-factor of a pressurized circular pipe is

$$C = 3.551 \times Q \times D^{-2.63} \times L^{0.54} \times H^{-0.54} \tag{3-2}$$

Where:

C = C-factor coefficient
Q = Rate of flow in gpm
D = Diameter of pipe in in.
L = Length of pipe in ft
H = Head loss in ft

C-factor coefficients are fairly constant within normal velocity ranges. Field results are verified by measuring head losses at two different flow rates and comparing resulting C-factors. C-factors should agree within 10 percent. Another check consists of running two test trials on the same pipe but using different equipment for each trial.

Usually, flow must be artificially induced to produce velocities and head losses high enough for achieving accurate measurements. In pipes 16 in. (406 mm) and smaller, high flow rates are induced by opening one or more hydrants. In larger pipes, it is necessary to use other methods, such as opening blowoffs, throttling parallel mains, operating pumps, or filling tanks. It is necessary to close valves to force the measured flow to pass through the pipe section where head loss is measured.

Pipe diameters used in C-factor calculations are the same diameters used in the model, preferably actual diameter. The length of pipe used in calculating a C-factor is the distance along the pipe between the pressure measurement points. Tests that include more than one pipe diameter between pressure sensors should be avoided.

NOTE: in. × 25.4 = mm

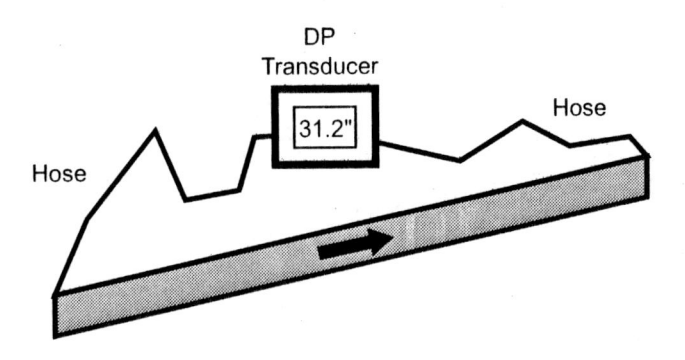

Figure 3-4 Parallel hose method for head loss

Head loss measurements are equivalent to the drop in the hydraulic grade line (HGL) between the inlet and outlet of a test section. Hydraulic grade line measurements are taken at hydrants, air valves, corporations, or other taps. Measurement techniques depend on field conditions and the amount of head loss being measured.

The Parallel Hose Method

The parallel hose method measures head loss using a hose or pipe connecting the ends of a test section to a differential pressure transducer or manometer, as shown in Figure 3-4. This method measures head loss directly without knowledge of elevation. Portable test equipment includes reels of 1/8-in. (3.2-mm) or 1/4-in. (6.4-mm) elastic tubing or rubber hose.

The parallel hose method generally requires at least 2 ft (0.6 m) of head loss to achieve accurate test results, provided differential pressure measurements are accurate within +/–0.1 ft (0.29 kPa). Pipe lengths and flow rates that produce measurable velocities and losses without exceeding the range of the transducer should be selected.

The parallel hose method is not suitable for testing some pipes, as the practical limit for the length of hose is approximately 3,000 ft (914 m). Some short test sections require extremely high flow rates to produce measurable head loss.

Sources of error include leaks and air in the hose. If possible, a valve should be closed immediately downstream of the test section and the section checked for zero head loss at zero flow.

The Gauge Method

The gauge method uses pressure gauges at the inlet and outlet of a test section. Pressure measurements along with known elevations define the HGL. Head loss is the HGL at the outlet subtracted from the HGL at the inlet, as shown in Figure 3-5. An accurate test requires a loss of at least 10 ft (3.1 m) using pressure gauges and elevations accurate to within +/–0.5 ft (0.2 psi) (1.4 kPa) of head. This level of accuracy requires gauge calibration with a dead-weight tester and accounting for the heights of instruments from benchmarks.

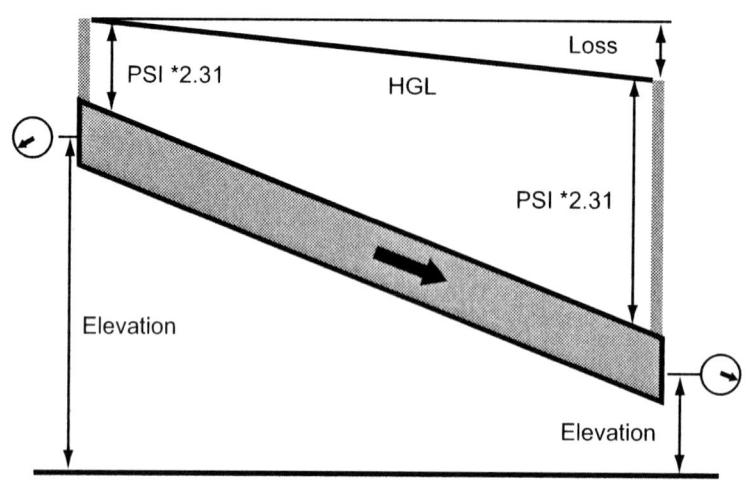

48 COMPUTER MODELING OF WATER DISTRIBUTION SYSTEMS

NOTE: psi × 6.895 = kPa

Figure 3-5 Gauge method for head loss

An advantage of the gauge method is that test lengths are increased to achieve additional head loss. However, isolating long test sections requires closing many valves. Another disadvantage is the need to know elevations at which the pressure gauge measurements were taken. If elevation data is not readily available, it is possible to close a valve downstream of the test section and record pressures at zero flow. The difference in static pressure is converted to feet of water and used as the elevation of the higher gauge, assuming the elevation of the lower gauge is zero or datum.

Other Methods

Other methods of measuring head loss or pressure are possible under certain field conditions. Reservoir levels define HGLs in some cases. The height to which water rises in clear hoses connected to pipe taps indicates the hydraulic grade line directly—this method is only practical when pressures are relatively low. Parallel pipes with no flow are used to transmit the HGL from one end of a test section to the other. Some methods of calculation use pressure measurements at several flow rates to eliminate the need to know elevations.

DIURNAL DEMAND MEASUREMENTS

Diurnal demand is the variation in water demand over a 24-hr period (see Figure 3-6). Knowledge of diurnal demand provides information on demand peaks, determines peaking factors, as well as the distribution of demand over time.

Peaking factors based on measured demand are actually preferable to assumed peaking factors, as climate, seasonal changes, and industrial activity can cause considerable variation in demand. Low peaking factors are typical in areas where large industries operate 24 hr per day and use water fairly steadily. Residential areas are associated with high peaking factors. Different subsystems have very different peaking factors even though they are supplied by the same basic distribution system.

NOTE: mgd × 378.5 = M m^3/$_{day}$ or mgd × 3.785 = mLd

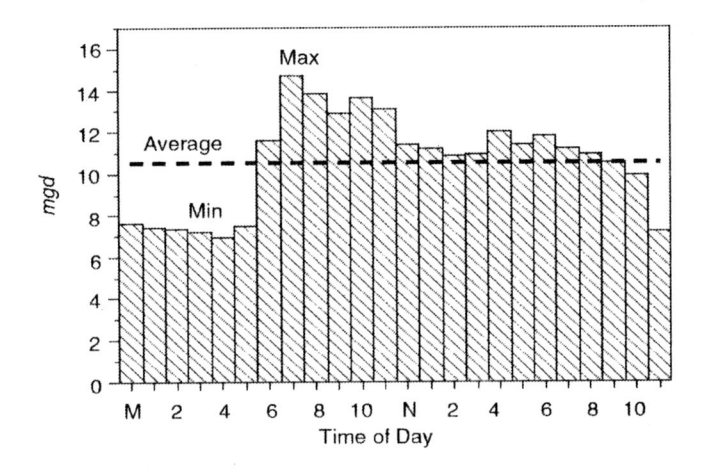

Figure 3-6 Diurnal demand measurement

Diurnal demand measurements also allow the user to easily check for leakage. Excluding industrial use, a minimum demand flow rate over 50 percent of the average flow rate indicates the likelihood of leakage. However, in warmer climates, automatic sprinkler systems are often set to water at night when other water demands are low, and when irrigation efficiencies are higher. The model needs to correctly interpret diurnal demand in order to identify leaks.

Diurnal demand measurements also allow the user to check demand distribution assumptions. Distributing demand from billing records requires allocation of water loss and involves splitting meter books between service levels. Demand distribution is based on census counts (demand per capita) or land use (demand per acre).

With demand measurements, the user checks the model against reality. For example, measurements may reveal that 30 percent of production flows into a district that only accounts for 20 percent of the total water sold. Further measurements may reveal faulty assumptions, accounting errors, leakage, open boundary valves between service levels, or faulty metering equipment.

Taking diurnal demand measurements involves isolating a particular region of a water system and monitoring hourly inflows and outflows for at least 24 hr. Districts are isolated by closing valves and taking advantage of service level boundaries. It is sometimes desirable to take measurements on an average day or a summer day (perhaps maximum day), depending on the intended use of the model. Flows in and out of tanks, service level boundaries, water treatment plants, and wells are measured. Typical flow measuring equipment includes source meters, pitot tubes, and tank water level recorders.

Use of tank water levels in calculating flow rates can introduce significant error, especially with tanks having a relatively large cross-sectional area. For tanks having large cross-sectional areas, high flow rates in or out of the tank result in only slight changes in water level, which are difficult to measure. Gauge resolution of 1 ft results in as much as 10 to 20 percent error in determining hourly demand for a small district. Multiple tanks compound the problem. Valving off as many tanks as possible reduces errors. Malfunctioning meters at plants and pump stations may result in large errors in hourly demand. Maintaining records of meter calibration data, and comparisons between meters and field data, is very useful to improve the accuracy of meter readings.

PUMP TESTS

Data from pump tests are entered into the model for simulating hydraulic performance and for calculating energy costs. Input data based strictly on design curves introduces significant errors, as true performance is sometimes quite different from design parameters caused by worn impellers, undocumented equipment changes, or trimming of impellers.

As shown in Figure 3-7, pump tests reveal how total dynamic head (TDH) and efficiency truly vary with flow. Most models generate pump curves based on three operating points—each operating point has a unique flow rate, TDH, and efficiency associated with it. Tests usually include normal conditions, design conditions, induced flow, throttled flow, and the shutoff head.

Master meters are often used for measuring flow, and the meters should be tested for accuracy, if necessary. Throttled flow and shutoff conditions are often achieved by closing a downstream valve. Induced flow is achieved by opening hydrants or blowoffs, shutting off other pumps, or filling a tank in the distribution system.

TDH is defined as the energy added by a pump per pound of water flowing through the pump. TDH measured in foot-pounds per pound is dimensionally equivalent to feet of water. The equation for the TDH added by a pump is

$$TDH = (P + V)_{OUT} - (P + V)_{IN} \qquad (3\text{-}3)$$

Where:

TDH = total dynamic head in ft ($\times$ 2.989 = kPa)

P = pressure head in ft

V = velocity head in ft

$_{OUT}$ refers to the pump discharge

$_{IN}$ refers to the pump suction

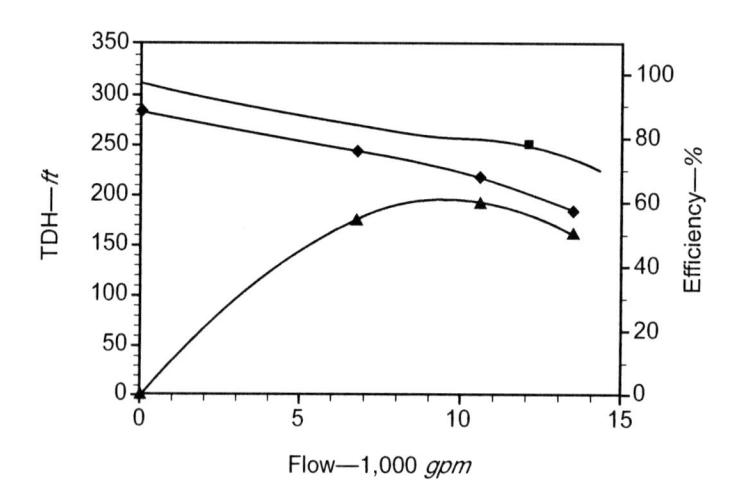

NOTE: ft × 2.989 = kPa
gpm × (6.3 × 10^{-5}) = m^3/s

Figure 3-7 Pump tests

Velocity head is computed by squaring velocity (in ft per second) and dividing by 64.4 ft per second squared, two times the acceleration of gravity. Velocities are calculated by dividing measured flow rate by the cross-sectional area of the pipe. Velocity heads are significant, especially for induced flows.

Some pumps have negative (less than atmospheric) inlet pressure, which increases the TDH as subtracting a negative value is equivalent to adding the value. For vertical turbine pumps submerged in reservoirs, inlet pressures and velocity heads are zero, and outlet pressures are referenced to reservoir water surfaces.

Efficiency is calculated by dividing the pump's output hydraulic power by the input electrical power to the drive motor. The equation is

$$E = 0.01885 \times Q \times TDH / KW \tag{3-4}$$

Where:

E = efficiency in percent
Q = rate of flow in gpm
TDH = the total dynamic head in ft
KW = electrical power in kilowatts

Electrical power is measured directly using portable test equipment. Electrical power is calculated from measurements of voltage, amperage, and power factor. Power factors must be measured to verify accuracy. Electrical power may also be calculated from the rotational speed of power meter disks, using the watt-hour per revolution factor (K_H) and multipliers stamped on most meters.

HYDRAULIC GRADIENT TESTS

Hydraulic gradient tests reveal how HGLs vary along the length of a pipeline. Data points along the HGL are computed by adding elevation to each measurement of pressure head, in feet. Hydraulic gradient tests provide data for calibrating computer models.

Hydraulic gradient tests consist of taking simultaneous flow and pressure measurements at intervals along a pipe path, under the condition of steady flow. In large systems, the tests follow trunk mains between plants, tanks, and pump stations, as shown in Figure 3-8. In small systems, the tests follow pipes from sources to flowing hydrants at key locations. Pressure measurements are taken at plants, on the inlet and outlet of pump stations, at tanks and major pipe intersections, and where pipe diameters transition. At key pipe sections, the flow rate is measured.

Pressure measurements in units of psig are converted to feet of water and added to pipe elevations to obtain HGLs. HGLs are then plotted against distance, as shown in Figure 3-9. The plots show the accumulation of head loss along pipe sections. The slope of the HGL is steep where there are significant hydraulic restrictions in the system.

To be truly useful, hydraulic gradient tests should be conducted under conditions of high flow when head loss is significant or when high flow can be induced. In small systems, a simple way to induce high flow is by opening hydrants. In large systems, it is not sufficient to open hydrants—the test is conducted when demand is high or when a tank is being filled.

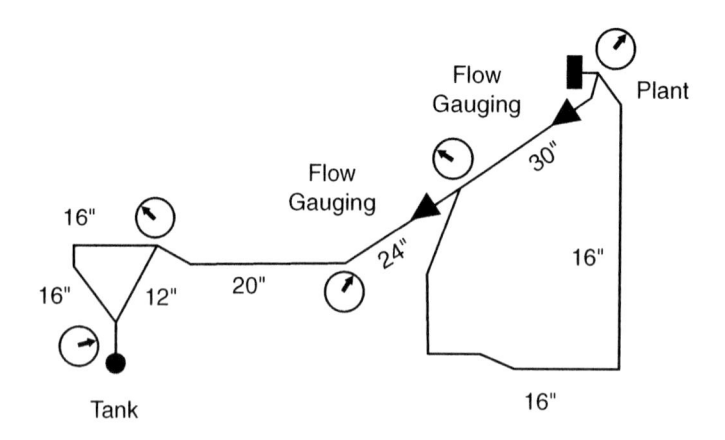

52 COMPUTER MODELING OF WATER DISTRIBUTION SYSTEMS

NOTE: in. × 25.4 = mm

Figure 3-8 Hydraulic gradient layout

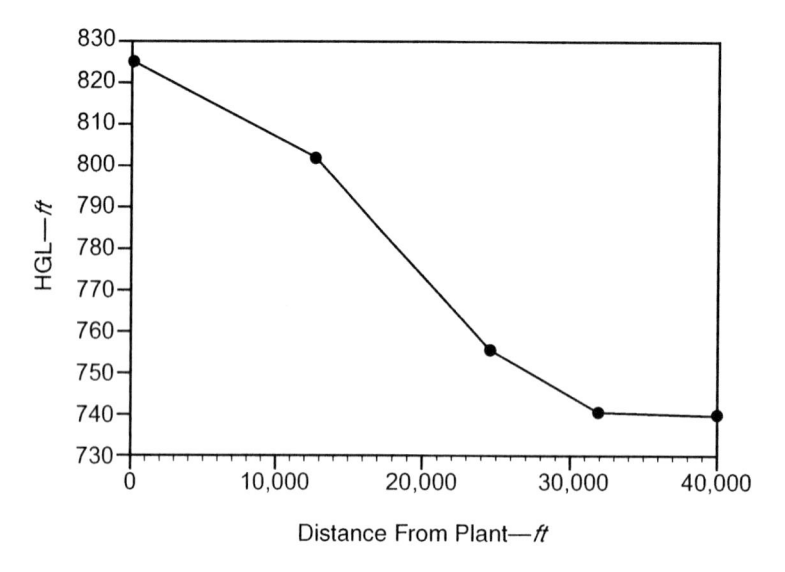

NOTE: HGL ft × 2.989 = kPa
distance ft × 0.3048 = m

Figure 3-9 Hydraulic gradient test

FIRE FLOW TESTS

Fire flow tests are used to determine friction factors in pipes near hydrants. These tests consist of simultaneously taking flow and pressure measurements at selected locations. These simple tests are an inexpensive way of checking a model against measured values.

First, static pressure measurements are taken under normal conditions. Next, flow is induced by opening a nearby hydrant, and dynamic pressure (residual pressure) measurements are taken under the condition of steady flow. In larger networks, or where larger mains serve the hydrant, two or more hydrants are opened to provide the necessary flows and resulting head losses. The flow rate is calculated from hydrant pitot gauge readings and the hydrant discharge port size.

To achieve sufficiently accurate test results, there must be a minimum 10 psi (68.95 kPa) pressure drop (static pressure minus residual pressure) and the pressure gauge should have a minimum 0.5 psi (3.5 kPa) resolution. Gauges are calibrated with a dead-weight tester.

Derived from the Hazen-Williams formula, the results are often expressed as flow rates available at 20 psig (137.9 kPa):

$$Q_A = Q_M \times \{(S-20)/(S-R)\}^{0.54} \qquad (3\text{-}5)$$

Where:

Q_A = the rate of flow available at 20 psig
Q_M = the rate of flow measured during the test
S = the static pressure in psig
R = the residual pressure in psig

gpm × (6.3 × 10⁻⁵) = m³/sec

Fire flow tests provide useful data for calibrating models of small systems. Modeling a measured flow and pressure is preferable to modeling the calculated flow at 20 psi (137.9 kPa) because of possible complications from check valves opening, flow reversals, or pump changes. For fire flow tests to be effective, critical boundary conditions, pump status, and reservoir levels also should be recorded.

Fire flow tests in large systems have only limited value in calibrating models, because most of the head loss occurs in the area near the flowing hydrant. Therefore, measurements from fire flow tests allow only limited calibration in the region of the hydrant—they do not provide a general indication of model accuracy.

OTHER TESTS

Some models require conducting other types of tests. Modeling pressure-reducing valves requires accurate downstream pressure measurements at known elevations. Pressure and flow measurements are useful in modeling other types of control valves. Pressure and tank level measurements are necessary for checking the accuracy of data from water plant charts and SCADA systems.

DATA QUALITY

Data collected for the purpose of model calibration must be more precise than data collected for routine operation and control. Calibration data will be used to adjust pipe roughness and demands. However, much of the time the head loss in water distribution systems is very small, and errors in the measurement of head loss are of the same order of magnitude as the head loss itself. (Walski 2000; Walski, Chase, and Savic 2001). Such data are useful in checking boundary heads and ground elevations but are useless for adjusting pipe roughness because at such low flow rates, head losses cannot be measured accurately.

The key to data collection for calibration is to ensure that
Head Loss >> error in measuring head loss.

This is achieved by minimizing errors in head loss measurements or increasing head loss. Head loss is maximized by taking measurements during peak demand periods or conducting fire hydrant flow tests. (Throttling valves should not be used because they introduce another unknown, the minor loss through a throttled valve.)

Head measurements consist of both pressure and elevation measurements. Pressures are read with a gauge calibrated to 1 psi (1 meter) or better. Elevation data should be at least as good (Walski 2001). While reading elevations from a USGS topographic map with a 20 ft (4 m) contour intervals are acceptable for model building, the elevation of calibration data points are significantly more accurate. Surveying the elevation of the gauge (not the elevation of the ground) is usually the most accurate way to determine if maps with 2 or 3 ft (1 m) accuracy are not available.

Adjusting demands and pipe roughness in models adjusts the slope of the HGL. Therefore, it is essential to know the boundary heads when calibration data are collected. Knowing that the pressure at a point in the system is 40 psi (275.8 kPa) without knowing the water level in nearby tanks, the operation and discharge head of nearby pumps, or the setting and status of nearby pressure-reducing valves renders such data useless for making calibration model adjustments. Relying on SCADA systems for this information helps if changes are gradual, such as changes in tank water level, but is not useful when pumps can turn on and off, and PRVs can open and close during a flow test because the polling interval of a SCADA system may miss a hydrant flow test. This requires stationing personnel at key points in the system during tests or installing data loggers to capture test results.

Readings through all flowmeters should be taken at a time corresponding to pressure measurements. It is very helpful to know the flow rate through meters during a hydrant flow test if it is suspected that flow rates change during the test.

REFERENCES

Bush, C.A., and J.G. Uber. 1998. Sampling Design Methods for Water Distribution Model Calibration. *Journal of Water Resources Planning and Management.* ASCE, 124:6:334–344.

Alden Research. 1995. *Calibration of an Insert Traversing Pitot Tube and Pitometer.* Computer Recorder. Alden Research Laboratory, Holden, Mass.

AWWA. 1998. Manual M31, *Distribution System Requirements for Fire Protection.* AWWA, Denver, Colo.

Strasser, A., Diallo, N., and E.J. Koval. 2000. Development and Calibration of Denver Water's Hydraulic Models for a Treated Water Study. In *Proceedings of the Annual Conference.* AWWA, Denver, Colo.

Walski, T.M. 2000. Model Calibration Data – The Good, The Bad, The Useless. *Jour. AWWA* 92.1.94–99.

Walski, T.M., Chase, D.V., and D.A. Savic. 2001. *Water Distribution Modeling.* Haestad Methods, Inc., Waterbury, Conn.

Walski, T.M., and S.J. O'Farrell. 1994. Head Loss Testing in Transmission Mains. *Jour. AWWA*, 86:7:62–65.

AWWA MANUAL M32

Chapter 4

Steady-State Simulation

INTRODUCTION

Hydraulic model simulations are divided into three different categories: steady state, time varying (also called extended-period simulation), and transient simulation. A steady-state model simulation predicts behavior in a water distribution system during a hypothetical condition where the effects of all changes in the operation and demands of the system have stopped. Steady-state analysis is sometimes compared with taking a snapshot picture of the distribution system. Of course, system never really reach a true steady-state condition, but the steady-state simulation is a very useful approximation of many network conditions. The steady-state assumption simplifies the analysis of a water distribution system. Steady-state simulations are often used to solve infrastructure-related design problems.

Extended-period simulations are a series of steady-state calculations linked together to approximate the behavior of a system over one or more days. Extended-period simulations are useful to model the change in pump operation, valve operation, tank levels, and water quality parameters. Extended-period simulations are often used for operational studies and are discussed in chapter 5. Water quality simulations are covered in chapter 6. Hydraulic transient simulations are used to predict very rapid pressure changes that are a result of pump failure events or sudden changes in valve position. These simulations usually cover a period of seconds or minutes. Hydraulic transient simulations are not covered in this text.

Many hydraulic problems in a water distribution system are solved by a steady-state simulation. These include sizing mains, pump stations, reservoirs, and planning the layout of a water distribution system as it evolves. This chapter details the following general topics:

- *Steady-state model calibration.* The model is tested to ensure that it predicts distribution system behavior with sufficient accuracy.

- *Selecting limiting conditions for design scenarios.* The modeler determines the design conditions under which the model predicts distribution system performance.

- *Establishing design criteria.* A model compares distribution system performance against a set of performance criteria to determine the adequacy of the system.

- *Developing system improvements.* The process for several categories of model simulations are described to explain the process of steady-state hydraulic modeling.

The procedures described in this chapter are the hypothetical and general case. Because general cases never strictly occur, it is important that the modeler is knowledgeable in hydraulics and applies sound engineering judgment to the procedures. System analysis is partly science and partly interactive trial and error that incorporates the feel of the distribution system. This chapter is intended as a general guide, pointing out important considerations and providing guidance on the scientific principles used in system analysis. The modeler uses this information as a base and tailors the analysis to site-specific information.

STEADY-STATE CALIBRATION

The first step in the calibration process is often a steady-state simulation.

Calibration is the process of fine-tuning a model until it simulates field conditions within acceptable limits. The traditional procedure involves trial and error. Newer calibration techniques include automated optimization using genetic algorithms.

Fine-tuning a model means correcting errors and adjusting the data to within reasonable limits. Examples of errors that are in the data include incorrect pipe sizes or interconnections that do not exist. Adjustments to the data usually involve demand distribution, friction coefficients, and controls, such as pump curves. Unreasonable adjustments, such as changing C-factors to 25 for lined pipes, may force models to match measurements without finding underlying problems, such as closed valves.

Field conditions for steady-state calibrations should include both flow and HGL comparisons, as shown in Figures 4-1 and 4-2. Matching HGLs without checking flows is not acceptable and vice versa. Compensating errors cause incorrect flows to produce correct pressures and apparent calibration for a certain condition, but flows for other conditions produce large pressure errors. Checking several conditions is good practice.

Acceptable limits of accuracy depend on how a model is used and the questions to be answered (Walski). For example, models used to design elevated tanks must predict tank HGLs well within 20 ft (6 m), because 20 ft (6 m) is the difference between an empty tank and a half-full tank. Better goals are HGLs within 5 ft (1.5 m) and flows within 10 percent. Calibration guidelines have not yet been adopted.

Calibrating a steady-state model is a trial and error process. Large models include many elements and finding those that need fine tuning may require several attempts. Calibration is like adjusting a television with many controls. Each control has a limited range of adjustment and some controls affect others. Genetic algorithms can find the optimum adjustments without trial and error. Tracer studies and water quality simulations are also useful calibration tools.

Testing Assumptions

The need for calibration arises because of uncertainties and assumptions in the data. Assumptions include Hazen-Williams coefficients, skeletonization, pump performance, and demand loading (Cruickshank and Long).

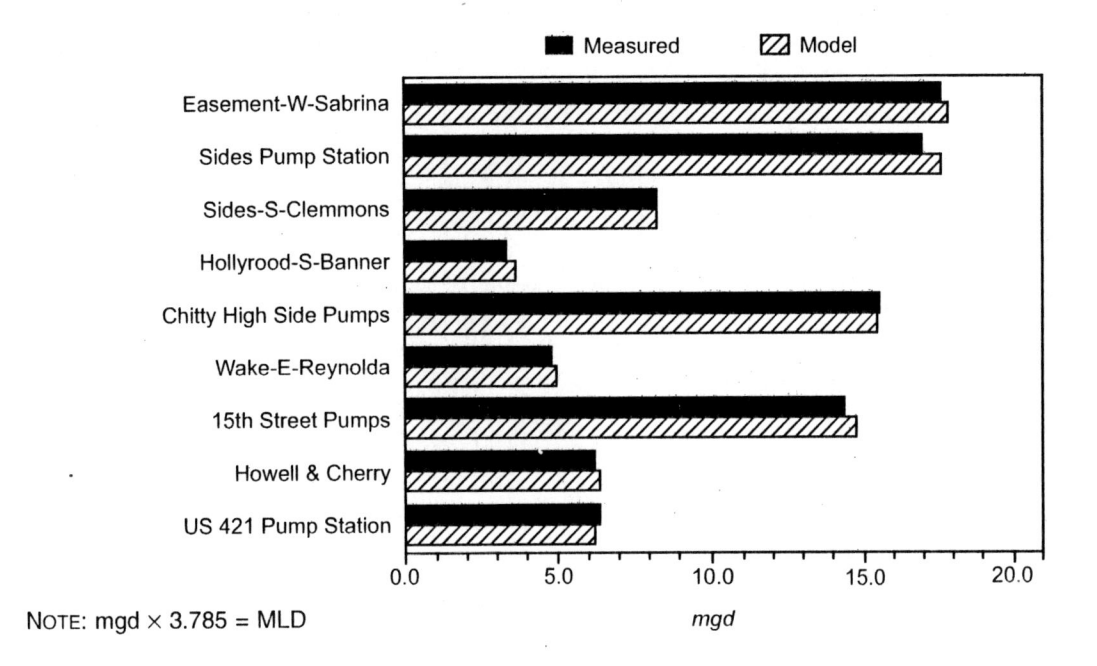

Figure 4-1 Flow calibration

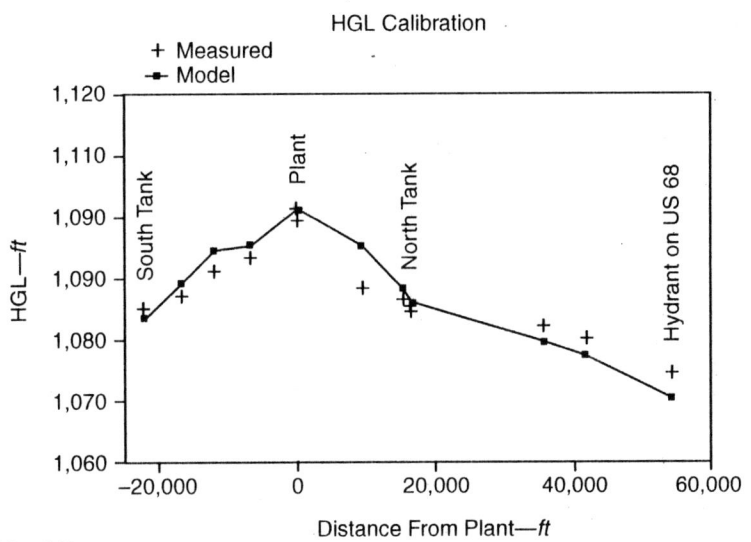

Figure 4-2 HGL calibration

Hazen-Williams coefficients are frequently based on pipe ages. Coefficients vary over a wide range and their correlation with age varies from system to system. In some cases, C-factors may be related to how far a pipe is from a source that deposits a chemical coating. Calibration verifies assumed coefficients.

Skeletonization is the practice of reducing the size and complexity of a model by leaving out small pipes. Calibration verifies decisions about what to leave in and what to leave out.

Models often simulate pump performance based on assumptions. Most models use pump curves generated from several input data points. The data points are often from manufacturer's design curves, which show performance for a new impeller. In reality, the impeller may be worn, or it may have been trimmed over the years. Calibration verifies assumed pump performance.

Calibration also tests assumptions involving demand. These assumptions include the geographical distribution of demand, the allocation of unaccounted-for water, and daily and hourly variation. For example, a constant demand factor may have been assumed throughout a system. In reality, demand peaks during business hours in industrial districts and during the evening in residential districts. Calibration reveals inaccuracies caused by bad assumptions built into the demand data.

Stressing the System

Steady-state calibration involves comparing measurements to predictions for a certain condition. The condition is defined by the demand in the system, tank levels, and the pumps that are operating. An important step in a steady-state calibration is choosing the condition to simulate. A stressed condition with high head losses produces more meaningful comparisons between measurements and predictions. Matching predicted HGLs to measurements proves nothing when head losses are low. Inducing high flow rates makes head loss large relative to measurement error and adjustments become more obvious.

Stressing small systems is relatively easy to do using fire hydrants. Flow from hydrants causes high flow rates and measurable head loss. Opening hydrants also provides an opportunity for flow measurements.

Stressing large systems requires more effort. Hydrants produce only localized effects. A better method is conducting measurements during the summer when demand is highest. Another technique is to simulate high demand by filling tanks and operating large pumps.

Operating conditions should be recorded carefully during all field measurements. Operating conditions include the date, time, tank levels, control valve settings, and number of pumps running. The model should duplicate these conditions.

Simplifying Initial Simulations

Sometimes a model is calibrated by temporarily simplifying it by removing unknowns until some of the calibration parameters are set.

The most useful simplification is setting as many correct flow rates as possible. Setting correct flows at inputs and pumps makes flows throughout the model more correct. Correct flows make HGL comparisons more meaningful and adjustments more obvious. For example, if the flow at a source is correct but the simulated HGL is too high, the model needs adjustments, such as higher C-factors, to reduce friction. Setting correct flows at booster pumps prevents problems in one service level from affecting other levels.

Methods of setting flow rates include eliminating pump curves and assigning negative demands at input points. A booster pump is modeled as a closed valve between two nodes, with the correct flow rate as a positive demand at the upstream node and a negative demand at the downstream node.

Another simplifying technique is shutting off some of the elevated tanks in systems with more than one tank. The model calculates the HGL where a tank "floats" with no flow in or out. This HGL is a useful benchmark—the tank fills at lower HGLs and empties at higher HGLs.

After a simplified model matches measurements, refinements should be added one at a time. The data for the refinements is the only source of error, making the necessary adjustments more obvious.

Plotting Hydraulic Gradients

Plotting hydraulic gradients helps to visualize the process of calibrating a steady-state model. A hydraulic gradient is a graph that shows the accumulation of head loss with distance, as shown in Figure 4-2. One axis is distance and the other is HGL, which is elevation plus pressure in ft of water.

Hydraulic gradients for calibration show the model predictions and actual measurements on the same graph. The graphs show how far HGL predictions are from measurements and where the discrepancies occur.

Hydraulic gradients for calibrating a model follow pipes from sources to important elements in the model. In small water systems, hydraulic gradients follow pipes from sources or tanks to fire hydrants used to induce flow. In large water systems, hydraulic gradients follow trunk mains from plants to pump stations and tanks.

Problems to Avoid

Several problems affect field measurements. These include inaccurate flowmeters and pressure gauges, velocity head at pumps, and bad assumptions based on pump nameplates.

Flowmeters at water plants and booster pumps are sometimes either fast or slow. Errors over 10 percent are not unusual. Pressure gauges are sometimes off by 10 or 20 psi (69 or 138 kPa), especially old gauges on pumps. Flowmeters are checked against a test meter or other measuring device. Gauges should be checked using a dead weight tester. HGL calculations should include gauge corrections and the height of gauges above benchmarks.

Velocity heads cause errors near pumps. The measured pressure at a pump flange is significantly lower than the pressure in larger, adjoining pipes. Most of the velocity head at a pump flange is converted to pressure in larger pipes leaving a pump station. Comparing a pressure measurement at a 4-in. (102-mm) pump flange to a model node on a 24-in. (610-mm) pipe causes an error of 10 ft (69 kPa) at a flow of 1,000 gpm (6.3×10^{-3} m^3/sec). Before comparing measurements to model results, velocity heads should be added to pressure measurements.

Another potential problem is bad assumptions based on pump nameplates. A pump may operate at conditions quite different from the nameplate rating, or the impeller may be worn. Two pumps with the same flow and TDH on the nameplate can have very different curves.

Sensitivity Analysis

After the model is calibrated to establish the limits of accuracy, or level of confidence, a sensitivity analysis is performed. Although sensitivity analysis is not essential, it provides greater confidence in the model predictions and a better understanding of the model's response to variations in key parameters. The sensitivity analysis addresses variations in the following system features:

- Increase in pipe roughness (typically 10–20 percent)

- Decrease in pipe roughness (typically 10–20 percent)

- Increase in demand conditions (typically 15–25 percent)
- Location of new large demands at several locations throughout the system
- Assessment of the model's sensitivity to storage tank levels and pumping options

SELECTING LIMITING CONDITIONS FOR DESIGN SCENARIOS

If a distribution system operates satisfactorily under the most severe demand conditions, it should operate satisfactorily for all conditions. For this reason, the most limiting demand conditions should be established and simulated with the model.

The diurnal demand curve is used to determine the limiting demand conditions. The curve for the maximum day condition is compiled and the total flow for the day averaged to determine the average maximum day demand. Figure 4-3 illustrates this curve. Points A and B represent two times during the day when the instantaneous demand equals the production rate from the supply sources. Point D is the peak hour flow rate, and Point C is the minimum hour flow rate.

Water treatment plants perform ideally when their flow is maintained at a constant production rate. This is why the production rate is set to the average maximum daily demand. The difference between the diurnal demand curve and the average maximum day at any point in time is the flow into or out of the storage facilities. If the available production rate is below the average demand for the maximum day, then the modeler must estimate the duration of maximum day conditions to ensure that the storage reservoirs are not emptied over a number of days. At the minimum hour demand rate (Point C), the demand for storage replenishment is at its maximum. This is often a limiting condition that must be analyzed to determine whether the distribution system can fill the storage tanks.

At the peak hour demand rate (Point D), the flow out of the reservoirs is at its maximum. The storage reservoirs, pumping facilities, and pipelines must be assessed at this condition to ensure that the customers have sufficient pressure.

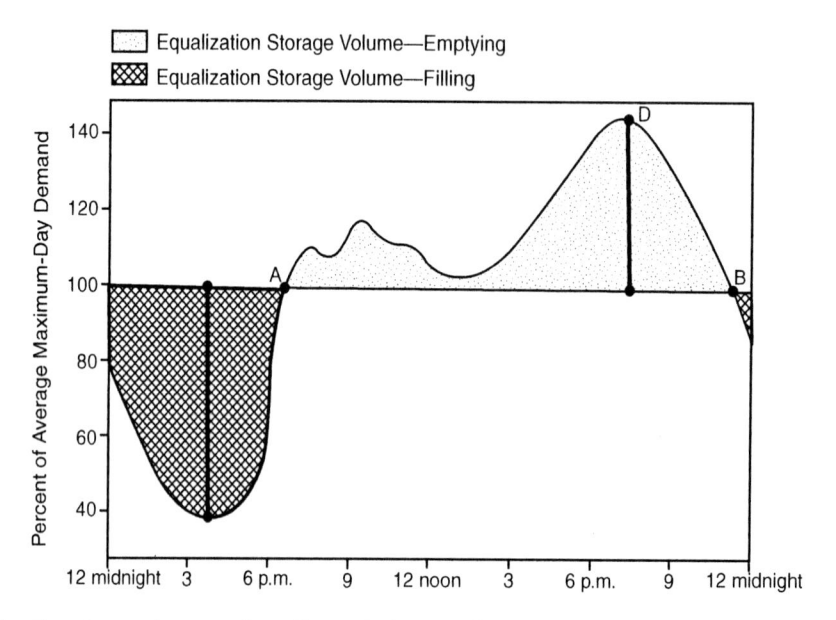

Figure 4-3 Idealized maximum-day diurnal demand curve

An important limiting demand not shown on the diurnal demand curve is the required fire flow demand. These demands are community specific and should be assessed as an additional flow superimposed on the average maximum day demand. Referring to Figure 4-3, this condition corresponds to a fire at Point B on the curve. Point B is the most limiting, as the reservoir storage available is at its minimum because of the higher demands over the previous hours.

The most common steady-state scenarios are average day, maximum day, maximum hour of the maximum day, the minimum hour of the maximum day, and maximum day plus fire. The choice of which condition to model depends on which question the modeler is trying to answer. A common approach is to use steady-state simulations initially to evaluate the system and work out the conceptual design of improvements, and then use extended-period simulations to check the final design. Table 4-1 summarizes some typical analysis types and the corresponding condition used to simulate the network.

It is not necessary to build individual models of each of these scenarios. Rather the modeler should develop a database containing the information required to construct the model and use peaking factors to change the demands to reflect the different scenarios. Typically, the base model is set up to reflect either the average day conditions or the maximum day conditions. Average day has the advantage of being able to correlate total system demands to the information from the billing systems. Maximum day is advantageous because it is easily adjusted to the most common design scenarios, either maximum hour or fire simulations.

Separate data files should be maintained for what-if analyses and for future growth scenarios. This avoids the potential of corrupting the model representing the existing system.

Calibration data should be collected on a regular basis and compared to the model output to ensure that the models continue to accurately represent the true operation of the system.

The following sections discuss the major demand scenarios in more detail to assist the modeler in developing the model.

Maximum Day

Maximum day is the highest average demand for a 24-hr period. Maximum day is a critical condition because it is the largest demand to be supplied completely from production, without using storage.

Storage cannot supply maximum day demand because tanks do not produce water. Any water supplied by tanks must be replaced within a 24-hr period. Otherwise, tank levels would drop from one day to the next. Eventually, storage could not provide water for peak hours or fires on subsequent days when demand is likely to be nearly as high as the maximum day.

Table 4-1 Typical model scenarios

Purpose of the Analysis	Recommended Steady-State Demand Scenario
Studies of normal operation	Maximum day
Production and pumping requirements	Maximum day
Design—small systems	Maximum day plus fire flow
Design—large systems	Maximum hour of maximum day
Tank filling capabilities	Minimum hour of maximum day
System reliability during emergency or planned shutdown	Condition when the emergency or shutdown is likely to occur
Model calibration	Condition during time when measurements were collected

Maximum day analysis is a two-step process for systems with several elevated tanks. An initial simulation calculates HGLs that correspond to zero flow at the tanks. These HGLs are used to fix tank levels in the final simulation.

The initial simulation involves fixing input flow rates using negative demands, constant flow devices, or pump combinations known to produce the correct input. Input rates should be equal to or less than the production capacities of sources. The total supply should equal the total maximum day demand.

The initial simulation also involves turning off all except one tank in each service level to prevent flow to or from storage. One tank in each subsystem is set half full by modeling a fixed HGL. The model calculates flow in or out at fixed HGLs, but in this case, the flow must be zero because supply equals demand.

Setting a tank half full reserves the lower half for fire protection and emergencies. Half full is the minimum tank level in the evening of the maximum day when equalizing storage is depleted.

Deciding which tank to set half full is seldom difficult. Operators or automatic control systems usually start or stop pumps based on the water level in a certain tank, often the tank farthest from the source.

HGLs calculated in the initial simulation provide useful information. The calculated HGLs at elevated tanks show how the tanks float relative to each other. Calculated tank HGLs are compared to overflow elevations. HGLs at sources are used to calculate TDH requirements for pumps. The initial simulation provides information for setting up a realistic final simulation.

The final maximum day simulation uses pump curves at sources and fixed HGLs at all elevated tanks. Pump curves are selected to approximate the input flows and HGLs from the initial simulation. The fixed HGLs at the elevated tanks produce little flow in or out, because supply equals demand and the tank levels were balanced for zero flow. The final maximum day simulation is a good general condition for a database because it is adjusted easily to either a fire flow or maximum hour simulation.

Maximum Hour

Maximum hour is the design condition for large water systems. Maximum hour usually exceeds maximum day plus fire in large systems with a maximum day greater than 10 mgd (38 MLD).

Maximum hour demand is modeled using peaking factors or set directly in the demand data. A maximum day database is adjusted to the maximum hour using a diurnal curve or peaking factors. Peaking factors are obtained from measurements or SCADA records.

Ideally, pumping rates at sources are the same as those in the maximum day simulation. For systems that use clearwells as equalizing storage, pumping rates for the maximum hour are higher than the average pumping rate for the maximum day.

Elevated tanks are simulated as fixed HGLs or flow inputs. Flow inputs are useful if some tanks, such as standpipes, have cross-sectional areas much different from others. Fixed HGLs allow the model to calculate tank input rates. The same fixed HGLs from the maximum day simulations are reasonable for the maximum hour. The maximum day simulation uses balanced tank levels at the lower limit of the daily cycle of emptying and filling. Maximum hour usually occurs in the evening of the maximum day when tank levels are lowest.

Ground storage that does not float must be pumped into the system. Maximum hour simulations use flow inputs or pump curves to utilize ground storage.

Tank depletions during the maximum hour should be checked against tank capacities. A rule of thumb is that a tank supplies a rate in mgd equal to twice its volume in mg, which assumes using half the volume in 6 hr. Another approach is multiplying depletion rates by an equivalent emptying time. Calculating the equivalent emptying time involves analyzing the 24-hr demand curve. Table 4-2 analyzes an actual demand curve by determining the equalizing volume requirement, shown in Figure 4-4, and calculates the equivalent emptying and filling times for a simple block demand curve, shown in Figure 4-5. The block curve consists of only three demand rates: minimum, average, and maximum.

Table 4-2 Calculation of equivalent emptying and filling times

Time of Day		Demand mgd	Supply mgd	Rate to Storage mgd	Rate From Storage mgd
Midnight	–1 a.m.	13.80	19.28	5.48	
1	–2	11.60	19.28	7.68	
2	–3	9.80	19.28	9.48	
3	–4	9.60	19.28	9.68	
4	–5	9.90	19.28	9.38	
5	–6	11.40	19.28	7.88	
6	–7	16.50	19.28	2.78	
7	–8	23.80	19.28		4.52
8	–9	21.30	19.28		2.02
9	–10	20.90	19.28		1.62
10	–11	20.50	19.28		1.22
11	–Noon	21.70	19.28		2.42
Noon	–1 p.m.	20.60	19.28		1.32
1	–2	18.50	19.28	0.78	
2	–3	17.80	19.28	1.48	
3	–4	18.10	19.28	1.18	
4	–5	17.60	19.28	1.68	
5	–6	18.80	19.28	0.48	
6	–7	24.40	19.28		5.12
7	–8	28.70	19.28		9.42
8	–9	30.30	19.28		11.02
9	–10	30.80	19.28		11.52
10	–11	27.50	19.28		8.22
11	–Midnight	18.80	19.28	0.48	
Totals		462.70	462.70	58.43	58.43
Averages		19.28	19.28	2.43	2.43

$mgd \times 3.785 = MLD$, $MG \times 3.785 = ML$

Calculation of Equivalent Filling Time:	
Actual Equalizing Volume	2.43 MG
Maximum Rate to Storage	9.68 mgd
Equivalent Filling Time	0.25 days (6.04 hr)
Calculation of Equivalent Emptying Time:	
Actual Equalizing Volume	2.43 MG
Maximum Rate From Storage	11.52 mgd
Equivalent Filling Time	0.21 days (5.07 hr)

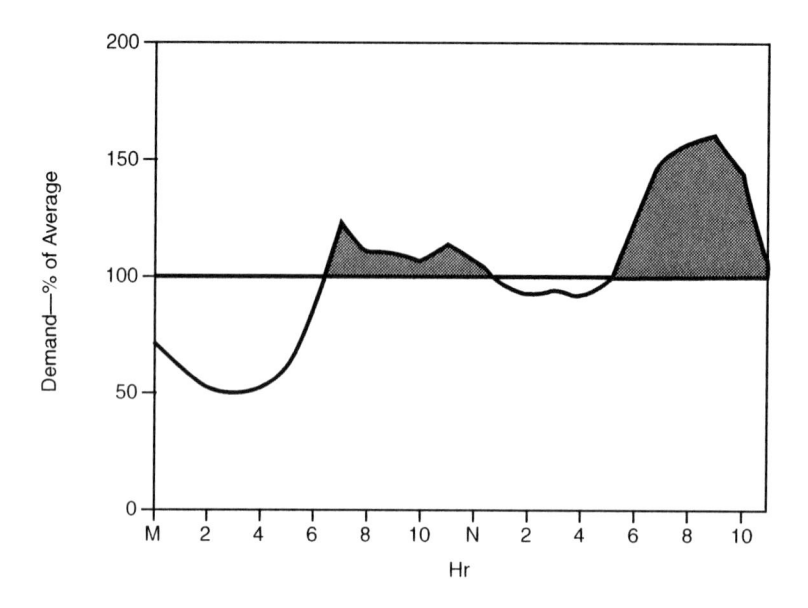

Figure 4-4   Equalizing volume requirement

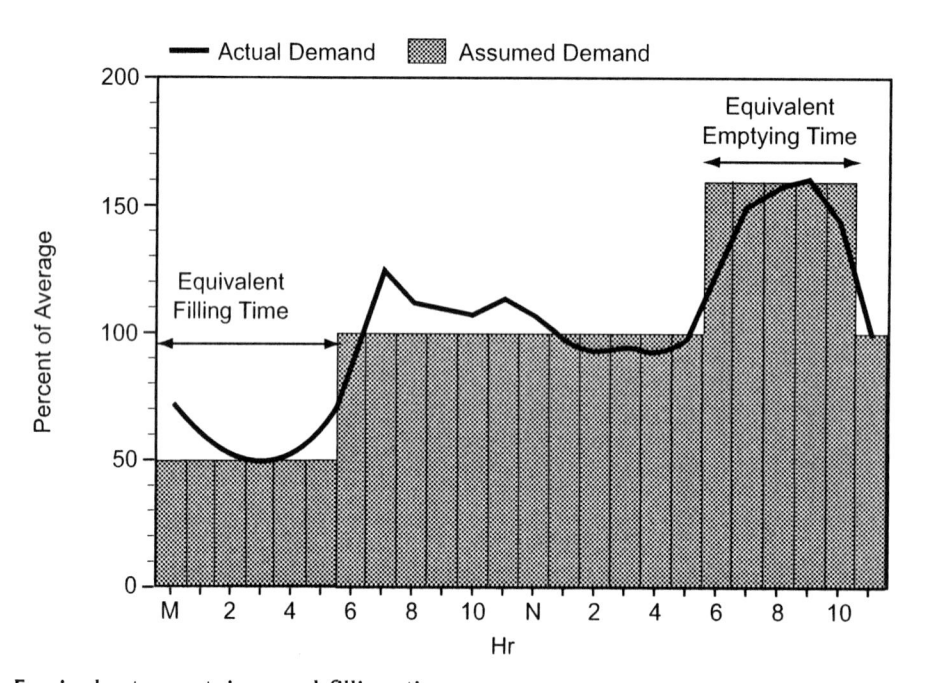

Figure 4-5   Equivalent emptying and filling times

Pressures are a key result of maximum hour simulations. Comparing pressures for various alternatives simplifies decisions for proposed improvements. If the model predicts pressures that are not within design criteria, improvements should be added or other changes should be made.

The HGLs at nodes with low pressures are significant. If the HGL at a low pressure node is within a few feet of the water level at the nearest tank, high ground elevations caused the low pressure. An HGL far below the nearest tank level indicates a problem caused by friction loss.

Average Day Simulations

Average day simulations model the normal operation of a system. Pump selections are optimized. Calculated flows are compared to annual average production and billing records. Steady-state simulations of average day create extended-period simulations that study power costs or water quality. Demand is adjusted to the annual average production using peaking factors.

The initial simulation for an average day uses the same procedure as the maximum day. Initial simulations use flow inputs equal to demand, with no flow in or out of tanks. Input flow rates are modeled as negative demand or control flow valves that input the annual average production.

Elevated tanks should be turned off to prevent flow in or out, except in each subsystem, one tank remains open with the water level fixed at a reference level, such as three-quarters full or half full, depending on how the system operates. The model calculates zero flow at the open tank because supply equals demand.

The final simulation uses the results of the initial simulation to model sources as pump curves and elevated tanks as fixed HGLs. The fixed HGLs at elevated tanks correspond to near zero flow in or out.

Model output from average day simulations are compared to design criteria for pump efficiencies and service pressures.

Fire Flows

Fire flow simulations are a common application of steady-state modeling. Fire simulations provide localized checks on model calibration. After simulating fire flow deficiencies, models are used to design improvements.

Three techniques are used to simulate fire flows. The first method is assigning a fire flow and calculating a residual pressure. The second technique is fixing the residual pressure and calculating a fire flow. The third method is imposing both flow and pressure constraints.

Assigning a fire flow is the same as adding demand at a node. If the node already has a demand, the fire flow is added after converting to consistent units and accounting for any peaking factors. The model then calculates residual pressures that are compared to the design criterion of 20-psi (138-kPa) (AWWA Manual M31 *Distribution System Requirements for Fire Protection*). The required flow is not available if calculated pressures are below 20 psi (138 kPa). The lowest pressure is not necessarily at the node with the fire flow. It could be at a nearby node that is higher in elevation.

The second method, fixing a residual pressure, is the same as setting a hydraulic grade line at a tank. The HGL corresponding to 20 psi (138 kPa) is the ground elevation plus 46 ft (20 psi [138 kPa] 2.31 ft/psi). The model calculates an available flow rate for comparison to the required flow. Setting a 20 psi (138 kPa) residual at one node may cause pressures below 20 psi (138 kPa) at others, so all nodes should be checked.

The third method of simulating fire flows imposes both flow and pressure constraints. The flow constraint, usually 3,500 gpm or 5 mgd (0.22 m^3/sec or 19 MLD), prevents unreasonably high fire flows that empty tanks within a few minutes. The pressure constraint, usually 20 psi (138 kPa), prevents unacceptably low or negative pressures. The model determines the limiting condition. The model simulates either the limiting flow with a residual pressure higher than the constraint or the limiting pressure with flow less than the constraint.

Any type of fire flow simulation requires a node at the location of the fire. New node and pipe data should be added, if necessary. Fire flow tests are often done with multiple hydrants open at the same time. Node elevations are particularly important and must be accurate within 2 ft (0.6 m) to calculate residual pressures within 1 psi (7 kPa).

Fire simulations are simple to run using a database created for maximum-day conditions. Fire flows are added directly to the demand data. Pump curves allow the model to determine source input rates during the fire. The model automatically calculates depletion rates at elevated tanks modeled as fixed HGLs.

Tank levels from maximum day simulations are reasonable for fire flow simulations. Maximum day simulations usually set tanks at half full, the lowest point in the daily cycle of emptying and filling. The worst case is a fire that starts when tank levels are lowest. In this case, half-full tanks simulate conditions at the beginning of a fire when firefighting needs are highest. The more probable case is tank levels that are above half full when the fire starts and below half full when the fire ends, so half full approximates average levels during the fire.

Tank depletion rates from the model are multiplied by the following durations (see M31):

Fire Flow	Duration
2,500 gpm or less (0.16 m³/sec)	2 hr
3,000 to 3,500 gpm (0.18–0.22 m³/sec)	3 hr

Tank capacities are checked against the volume obtained by multiplying the fire duration by the tank depletion rates. If the depletion of any tank is more than half its volume, the tank may empty during the fire. The fire flow simulation is repeated with the tank HGL set just above the empty level and again with the tank turned off. If steady-state simulations indicate possible problems with empty tanks, extended-period simulations are warranted.

Replenishment Simulations

Replenishment simulations check the ability of the system to fill storage. This is a critical condition in some systems.

Demand is adjusted to the minimum hour using diurnal curves or peaking factors. Peaking factors are from measurements or SCADA data.

Ideally, pumping rates at sources are the same as those in the maximum day simulation. Some systems use higher pumping rates at night because energy costs are lower. Pumping rates are limited at plants trying to fill clearwells at night.

Elevated tank HGLs should be fixed at the overflow so the model can calculate filling rates. If some tanks fill much faster than others do, the simulation should be rerun assuming closed altitude valves on the tanks that filled fastest.

Ground storage tanks often fill at constant rates controlled by automatic valves. This situation is modeled as a pressure-sustaining valve. An alternative is assigning the required filling rate as demand at a node. In this case, the calculated HGL must be at or above the overflow; otherwise the fill valve is wide open and may not be able to maintain the required filling rate.

Filling rates calculated by the model are multiplied by an equivalent filling time. The resulting volumes are compared to depletions during maximum hour conditions. The equivalent fill time is determined from the 24-hr demand curve, as shown in

Figures 4-4 and 4-5. If steady-state simulations indicate problems filling tanks, an extended-period simulation is needed.

Model output should be checked for excessive pressures produced by fill conditions.

Emergency Conditions and Reliability

The limiting conditions previously discussed assume that all components of the system are operable. Water systems are also subject to emergency situations and planned maintenance activities in which certain components are no longer available to meet the water demands. Examples of emergency situations include earthquakes, well contamination, major power failures, or transmission main failures. Planned maintenance activities include water treatment plant shutdowns, reservoir cleaning or repairs, and transmission main valve repairs. System analyses assess the degree to which the system is relied on to function adequately when system components are inoperable. The reliability designed into a water distribution system is a policy decision that is based on a number of factors, including the frequency of the condition occurring, the contingency plans to mitigate the effects to the customers, and the cost to the utility.

Power failures and critical water main failures are situations that occur frequently, so the distribution system should be designed to maintain a minimum level of service to customers. The minimum level of service is defined in terms of a particular water demand scenario, such as "the average day demand should be met during a line break."

It is important to note that even though the system is capable of meeting the maximum hour criteria, a critical line break or a 6-hr power outage may debilitate the system to a greater degree. Analyzing the distribution system under emergency conditions requires engineering judgment. The important objectives of such analyses are to identify potentially vulnerable portions of the system and identify cost-effective solutions to improve the reliability. These alternatives are assessed against other utility policies to determine which approach best meets the needs of the utility in maintaining supply to the customers.

DESIGN CRITERIA AS ANALYSIS CONSIDERATIONS

Design criteria define system capabilities by specifying the performance requirements of the system components. Thus, whether the objective of the analysis is the design of a new system or improvements to an existing system, the design criteria define the potential solutions and are the standard against which system performance—both observed and predicted—is compared.

The computer model predicts the performance of the distribution system under various demand conditions. To identify the deficiencies, the model-predicted performance is compared to the established design and operational standards. Inadequate system pressures generally indicate deficiencies in a system. These deficiencies are caused by any of the model components including piping, pumping, and storage or caused by inaccuracies in the assumed system operating conditions.

Pressure Design Criteria

When assessing the adequacy of a system, the first parameter to check is the predicted pressure. There are generally three design pressures that are defined for each community: maximum pressure, minimum pressure during peak hour, and minimum pressure during a fire flow. Pressure fluctuations at a single point should also be kept to less than 20 or 30 psi (138 or 207 kPa).

The maximum pressure refers to the maximum pressure that customers experience. It is often in the range of 90–110 psi (620–759 kPa). It is important however to conform to any limits imposed by local building codes. In some communities the maximum is 80 psi (551 kPa) for internal plumbing. In these instances, the distribution system is sized for the higher pressure and individual pressure-reducing valves installed on the service lines to customers.

The minimum pressure during peak hour refers to minimum pressure at customers' taps during normal system operation. This value is typically in the range of 40–50 psi (276–345 kPa) and ensures that there is adequate pressure to the second story fixtures within a property if internal plumbing is configured correctly. Historically, minimum pressures as low as 30 psi (207 kPa) were used in the design of water distribution systems, however, when using this pressure the utility expects customer complaints because of the lower pressures. The newer plastic piping used within residential properties exacerbates the low pressure problems because of smaller internal diameters seen with this service line material. The minimum pressure also affects the design of lawn irrigation systems. The lower the pressure, the more piping and sprinkler heads required to ensure adequate coverage. It is important to note that in some communities where residential fire sprinkler systems are required by legislation, the minimum acceptable pressure is 50 psi (345 kPa) for them to operate properly. Backflow preventers are required in many commercial, office, industrial, and retail buildings. These devices cause a 5–15 psi (34–103 kPa) head loss, so minimum pressures should be set accordingly.

The recommended minimum pressure during fire flows is 20 psi (138 kPa). This value has been recommended by the National Fire Protection Association (NFPA) in the US and by the Fire Underwriters Survey (FUS) in Canada. The pressure is the minimum desired pressure in the distribution main closest to the flowing hydrant. The value of 20 psi (138 kPa) is used as it allows for adequate supply of water to the pumper vehicles while overcoming the friction losses in the hydrant branch, hydrant, and suction hoses to the pumper vehicle. This standard assumes typical hydrant branch configurations: if the hydrant branches are very short or very long, the standard is adjusted to reflect the decreased or increased losses because of friction in the hydrant branch. In addition to confirming the minimum 20 psi (138 kPa) at the hydrant location, the modeler should also be checking minimum pressures in other parts of the network to ensure that the system will remain sufficiently pressurized to prevent water main damage caused by collapse. Another consideration is the potential for backsiphonage of nonpotable water from customer properties or infiltration of groundwater around the pipes when very low pressures are seen in the water lines.

Depending on the extent of distribution system deficiencies, the distribution system model may predict negative pressures. Predicted negative pressures generally do not actually occur in the system. They appear in the model predictions because the mathematical model is forced to supply the full amount of all nodal demands. The significance of the predicted negative pressures is that the system demands cannot be met. In some cases, negative pressures actually occur in limited areas of the system, for example in high elevation areas. However, in general, a deficient distribution system simply fails to meet demands.

Piping System Design Criteria

In general, a distribution system is considered to have deficient pipe looping or sizing if the following conditions are seen:

- Velocities greater than 5 ft/sec (0.6 m/sec)

- Head losses greater than 6 ft/1,000 ft (18 kPa/300 kPa)
- Large pipe diameters (16 in. [406 mm] or greater) having head losses greater than 2 ft/1,000 ft (6 kPa/300 kPa)

Although none of these conditions are without exception, they are a concern to the modeler as they can indicate a waste of energy, which requires additional pumping. The modeled solution may have the least capital cost, but the operating costs are substantial over the life of the pipe. It is important to note that as velocity increases, pipe head losses increase exponentially. As velocities approach 10 ft/sec, the potential exists for problems to emerge, such as water hammer.

The design of new pipelines should address two basic features: sizing and routing. The size of the pipe is determined by the maximum flow rate carried, typically the maximum day plus fire flow condition for distribution mains. The maximum hour flow rate or the maximum storage replenishment rate is typically the maximum flow for the large transmission mains. The pipe size is selected to ensure that the velocity limits and head loss limits described above are not exceeded.

The second pipeline design feature to establish is routing. In general, looping of mains are used wherever possible. A looped system provides supply from two or more directions for large demands, such as fire flows, and provides backup and redundant supply paths in the event that a section of pipe is removed from service. The location of the new pipe is determined by the availability of right of way, construction easements, and common sense. When locating new pipelines, consideration should be given to the projected growth characteristics of the service area and where major demand areas are expected to be located. Routing studies consider the impact of development staging. Development restrictions may be required to limit properties needing large fire flows until a certain portion of the network and looping is completed.

Pumping Systems Design Criteria

Distribution systems have both high-service pumps and booster pumps. The difference between these pumps is that the high-service pumps obtain water from a reservoir open to atmosphere and booster pumps obtain their water from the already pressurized distribution system. Regardless of the pump type, the pumps must be sized to meet the full range of system demands, and all of the limiting demand conditions must be considered. This includes both the extreme flow conditions and the average conditions. Consideration of the duration of each condition should be made when selecting the appropriate pump size. The best choice may be a series of parallel pumps of varying sizes that can be operated to most efficiently match the pumping requirements.

In systems with elevated storage, constant speed pumps are used for high service and booster pumping, and storage is used to equalize the pumping rate over the range of water demands. For systems without elevated storage tanks or with insufficient equalization storage, it is desirable to use variable speed pumps to efficiently change the pumping rates to meet the system demands.

As indicated previously, high-service and booster pumps are installed as multiple pump installations. The pumping installation is designed to provide the maximum-day demand with any one of the pumps—preferably the largest—out of service. Individual pumps and combinations of pumps are sized to meet the range of demands from average day demand to maximum-day demand plus fire flow. When variable-speed pumps are required, it is not necessary to have all pumps as variable speed. However, for redundancy, the design should allow for maintenance of each pump individually while still maintaining the ability to vary pumping rates with system

70   COMPUTER MODELING OF WATER DISTRIBUTION SYSTEMS

demands. This is accomplished through the use of two variable-speed pumps per pumping station or, if multiple pumping stations supply the same pressure zone, one variable-speed pump per location.

A family of system head curves are developed for each pump station. These curves are combined to develop characteristic curves for the pump station based on the different pumping configurations. It is important to note that the pump curves provided by the pump manufacturer are field verified on a regular basis to ensure the pumps are performing as expected. A typical pump curve and a system head curve used to select the pump size are shown in Figure 2-2. The pump will operate where the two curves intersect, and the pump should be sized to operate at its highest efficiency.

The system head curve represents the required head for a range of flow rates. It is a relatively easy curve to develop for a single pipe. The system head curve is calculated by adding the static head, the head loss through the pipe, and the pressure required at the end of the pipe. However, the situation is more complex in a pipeline grid where water is conveyed to many points and via various routes. The hydraulic model is used to simulate the varying demand conditions, and the required system head curve is developed based on the pressure standards. This curve is plotted and the pump selection is made based on the percentage of time each condition is expected to occur. Figure 4-6 illustrates this selection process.

In selecting the pumps for a specific application, the efficiency of the pumps over the range of operating conditions must be evaluated. Figure 4-7 shows the typical efficiency range for a pump curve. The specific information on pump efficiency is available from the pump manufacturer. The modeler selects pumps that show the highest efficiencies over the expected ranges of demands. For multiple pump applications, the most efficient combination of pumps are determined and the operations plan established to maintain this sequencing order.

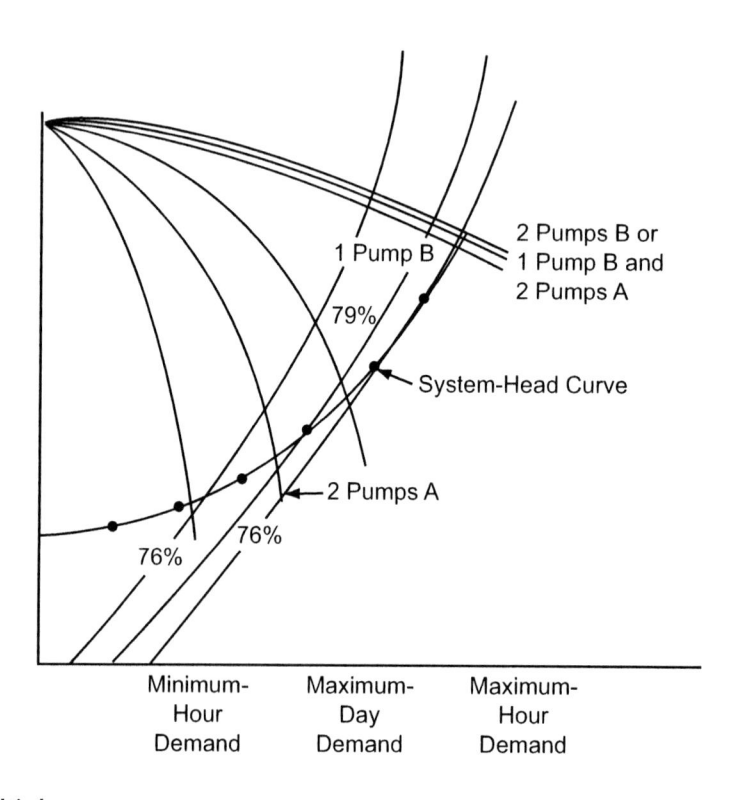

Figure 4-6   Multiple pump curves

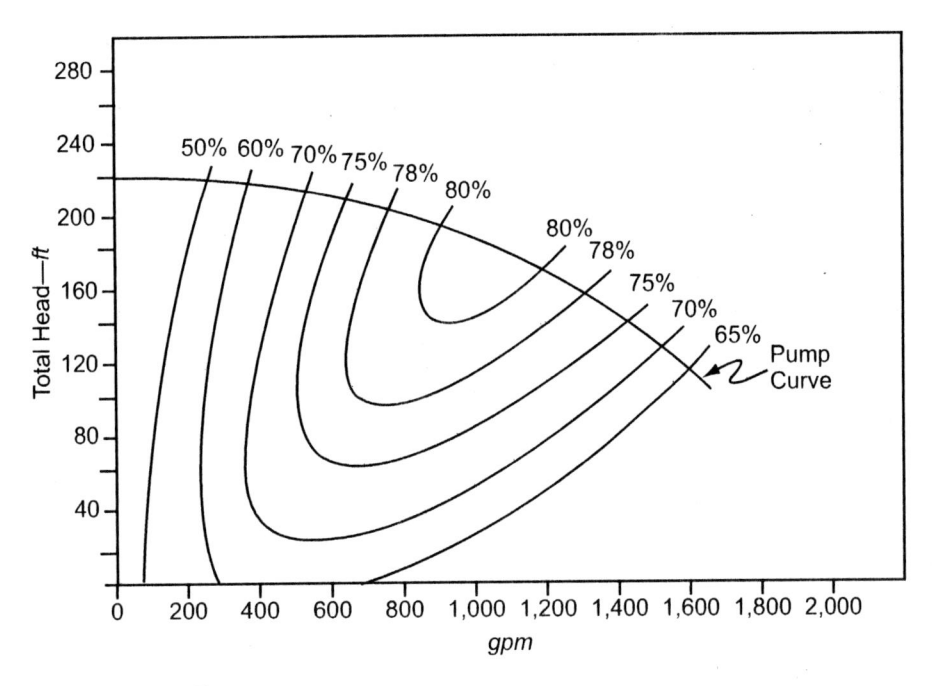

Figure 4-7 Pump curve efficiency

Energy Savings

Typically, pumping accounts for 75 percent of the power consumed by water utilities. Consequently, selecting the most efficient pumps for the range of demands and establishing the optimum sequential ordering of pumps has a substantial impact on the power consumption and cost to the utility.

In addition to selecting efficient pumps, careful scheduling helps reduce energy costs. To assess this potential, the electrical rate structure is analyzed to determine the least expensive times to pump. Electrical rates consist of two components, the demand charge and the energy charge. The demand charge is the maximum instantaneous power usage at a facility. In some rate structures, the demand charge is based on the highest demand in the last twelve months, so if the peak was an isolated event, the utility pays a substantial amount of money for the one occurrence. The energy charge represents the total amount of energy used over the billing period. The cost for energy varies at different times of the day, so if the utility schedules reservoir-filling activities during the lowest cost period, additional savings are found.

The electrical industry throughout North America is undergoing substantial changes as different states and provinces enact deregulation. The modeler should meet with power industry representatives to review the impact of different operating strategies on the power cost of the water utility. Generally, these different operating strategies require that the water utility has additional pumping and storage capabilities to allow the operation to take advantage of power cost savings.

Storage Facilities Design Criteria

Storage facility design criteria must address many interrelated factors, including system storage requirements, elevated versus ground storage and the number and location of the storage facilities. The volume of storage required is classified into three primary components: equalization storage, fire storage, and emergency storage.

Equalization Storage

Equalization storage is the amount of water required to meet demands in excess of the production and delivery capabilities. This storage is generally less expensive to provide than the production facilities, pumping, and piping that is required to meet all instantaneous demands. The amount of equalization storage maintained by a community should be determined based on a comparison of the production capabilities versus the demands expected on the system. In most communities, this is based on the maximum-day condition and ensuring that the equalization storage is sufficient to meet the demands that exceed the production capabilities. Typically, the equalizing storage requirement is 10 to 15 percent of the average demand over a 24-hr period for large systems, but equalizing storage could exceed 30 percent for small service areas or arid climates.

Fire Storage

The fire protection needs of a community typically are determined by The Insurance Service Office (ISO) in the US and by the Insurance Advisory Organization (IAO) in Canada. The requirements calculated by these organizations are based on the building types, land use, water supply facilities, and the response capabilities of the local fire department. The fire storage volume is determined by multiplying the required flow duration by the maximum fire flow in each service area of the distribution system. AWWA M31 includes tables showing the recommended duration for various fire flows. Factors, such as redundant piping and alternate supplies, are considered in determining the final requirements for the community.

Large fire flows may require more than one storage facility and a secondary piping route to the site. Typically, fire storage is obtained from reservoirs located within the same pressure zone as the fire.

Emergency Storage

Emergency storage provides water during events, such as pipeline failures, equipment failures, power outages, water treatment plant failures, raw water contamination events, or natural disasters. The amount of emergency storage is a policy decision based on an assessment of the risk of failures and the desired degree of system dependability. An assessment must be made of the type and nature of the emergency condition, including the frequency, intensity and dur,ation. In general, a vulnerability analysis, such as described in AWWA Manual M19, *Emergency Planning for Water Utility Management,* should be used to determine emergency storage requirements. In addition, some state or provincial regulations indicate the minimum emergency storage required for the community based on the average daily demands.

Storage Allocation

In evaluating system storage and developing operational plans for its use, the allocation of storage for equalization, fire, and emergency must be addressed. It is useful to consider storage schematically. Figure 4-8 shows how the three storage components are positioned in the reservoir. Storage for equalization occupies the top portion of the reservoir. During the day, this volume increases and decreases as demands are pulled from the reservoir when demand exceeds production rates. Fire storage is positioned in the reservoir, under the equalization storage. Emergency storage occupies the bottom portion of the reservoir. The importance of designating levels to the different storage components is to ensure that the volume has the correct hydraulic grade line for its intended purpose when dealing with elevated storage reservoirs. A fire may occur when the equalization storage has been completely consumed for the day, and the remaining hydraulic grade line must be sufficient to deliver the required flows at the proper pressures.

Elevated versus Ground Storage

The evaluation of the type of storage to construct includes high-level ground storage, elevated storage, and ground storage with pumping (see Figure 4-9). If there is high ground in or near the service area, ground storage connected directly to the system is used. However, for most distribution systems, the use of elevated storage, ground storage with pumps, or a combination of both must be evaluated.

There are advantages to both elevated storage and ground storage. Elevated storage maintains the system hydraulic gradeline without using pumps or controls, which results in a simpler operation. Elevated storage is considered more reliable than ground storage because it provides water during a power failure. Finally, elevated storage provides stable system pressures.

The major advantage for ground storage is the reduced construction costs per unit of volume. Ground storage is built to hold large volumes of water, whereas the size limit of elevated storage is about 3 MG (11 ML). Ground-storage facilities with pumps can be adapted to changing system hydraulics by changing pumps and controls. In contrast, elevated tanks must operate at the specific hydraulic grade line defined by their geometry.

Frequently, as water systems expand, the required hydraulic gradeline and water volumes change. This sometimes results in situations where either there is little turnover in the elevated reservoir because the system hydraulic grade line is higher than the tank, or conversely, the tanks deplete too quickly during high-demand periods. Ground storage facilities with pumps are regulated to respond to system demands, and their circulation ensured by strategically filling and pumping.

The operation and maintenance costs for steel elevated tanks are generally higher for painting and corrosion protection than ground storage. Power costs vary according to the conditions under which the facility is operated. Ground storage has power costs because of the associated pumping requirements. Elevated storage reduces power costs by allowing lower pumping rates during peak hours when power costs are highest.

Water freezing in elevated storage is a problem in northern climates where there are numerous days with temperatures below the freezing mark and limited ability to maintain water circulation.

Number and Location of Storage Facilities

The factors to consider when determining the number and location of storage facilities include the size of the reservoir, the location of other system components, hydraulic elevation, and fire flows.

The required number of reservoirs is determined using a trial-and-error analysis. The analysis determines the cost-effective number of storage reservoirs, taking into consideration the impact of cost, piping, pumping, and storage. A key factor to establish is the limiting size of each reservoir. The reservoir type (elevated vs. ground storage) determines the limiting factor. In general, the reservoir is as large as practical, but small enough to be filled within a reasonable amount of time. The use of individual reservoirs for each region to be supplied is compared against the cost of regional reservoirs supplying multiple areas. The cost comparison includes reservoir construction and operating costs, as well as the associated pumping and piping required for each alternative.

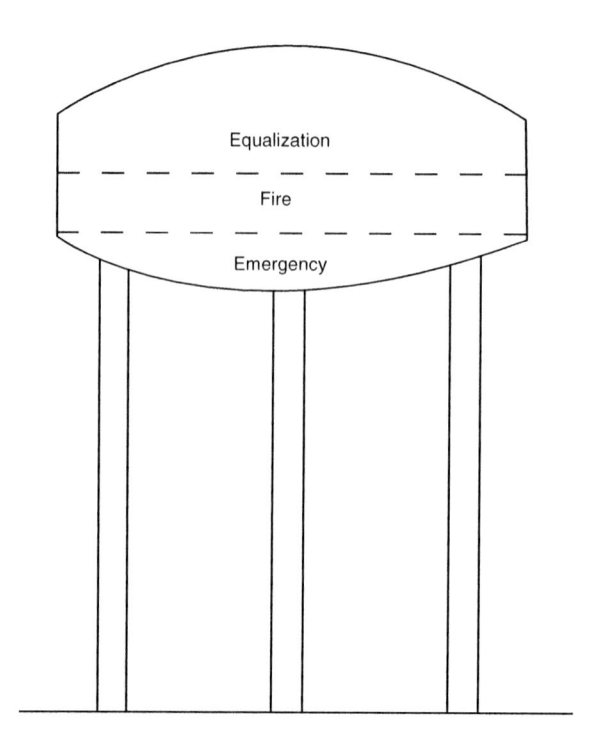

Figure 4-8 Storage allocation

Reservoir location is also an important design consideration. Reservoirs should be connected to major transmission mains, so that high flow rates are adequately transferred to and from them. Also, reservoir sites should be located at the extremities of the system opposite from the production source, allowing the reservoir to serve new growth occurring beyond the existing service area.

For effective fire flows, the storage facility should be located to provide coverage over the area, while avoiding unnecessary overlap with other storage sites. It is beneficial to locate the site as close as possible to the areas requiring the largest fire flows.

Water quality is another consideration. Large tanks with little turnover produce stale water and poor water quality. Reservoirs should be located and sized in a way that allows water to be circulated regularly.

DEVELOPING SYSTEM IMPROVEMENTS

A principle objective of distribution system modeling is to develop and evaluate potential system improvements. Comprehensive plans should be developed for system improvements that address the needs and deficiencies identified in the system analysis. The plans developed should include both an assessment of infrastructure requirements and an assessment of changing consumption patterns to identify the optimum solution. Some typical plans that utilities develop are listed in the following subsections.

In system modeling, a field-calibrated model is used to evaluate a distribution system for both existing and projected design scenarios. The design criteria are used to ensure that the proposed system meets the customer needs.

Before running each analysis, the modeler prepares an outline or index of model runs that describe each situation to be simulated. Each model run is either named or numbered, and a printout is maintained of the results of each analysis. Notes are taken on each system so that the evaluations are readily understood by anyone referring to them in the future.

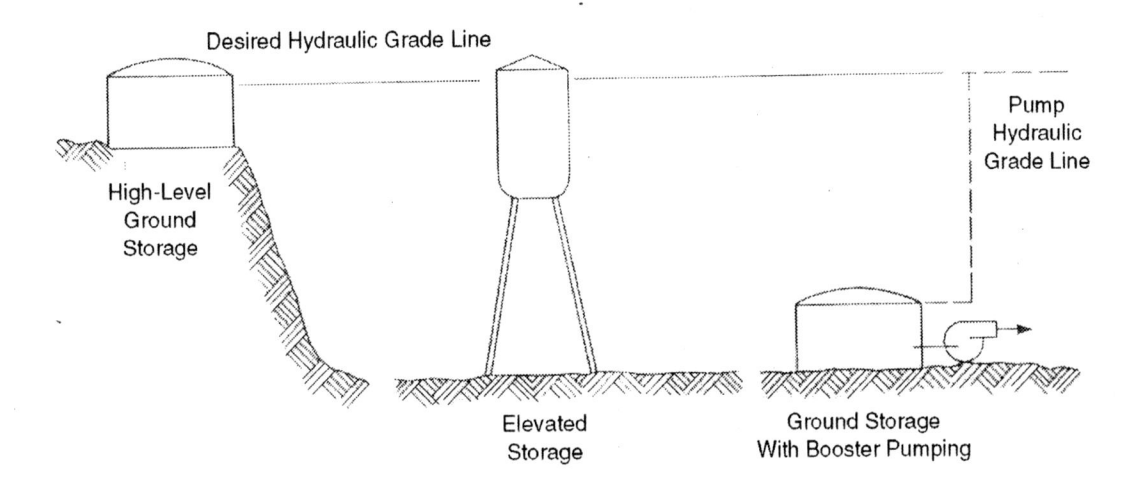

Figure 4-9 Types of storage and elevation

Master Planning

Master planning is defined as the development of long-range plans for the growth of the water distribution system. Typically master plans are developed for the 5-, 10- and 25-year projections. They include the assessment of the most cost-effective solutions for the location of new production or storage facilities and the expansion of the large transmission main network to support growth in the community. Frequently, master plans also analyze how to interconnect separate systems for emergencies.

Recently, water utilities were promoting the concept of Integrated Resource Planning. In terms of hydraulic system modeling and master planning, this means that the modeler looks not only at the construction and expansion of water infrastructure, but also assesses the impact of changes in customer demands caused by water conservation or shifting demands to other times of the day through customer education programs.

Master planning is increasingly done using all-main-models constructed from GIS databases, where the models are used for water quality and other kinds of detailed analyses.

Subdivision Planning

Subdivision planning is a localized analysis that considers the staging and growth patterns of new areas in the distribution system. The master planning model identifies the transmission main that supplies the volume of water required to the area from the production source. The subdivision model outlines the details of how the area should grow in a sequential manner.

The modeler considers the effects of staging from the perspective of system reliability by assessing the number of lots on a single feed. Fire flows are confirmed at each stage and, if insufficient flows are available, contingency plans developed to either limit high-density development or provide additional looping to safeguard property.

In some communities, adjacent developments can affect the ability of a neighborhood to grow in a cost-effective manner. Subdivision modeling is used by utilities to identify these problems and address them before they occur.

Rehabilitating Neighborhood Distribution Mains

Rehabilitating neighborhood distribution mains is similar to subdivision planning, as it deals with the design of water systems to individual customer lots. The primary difference is that because the rehabilitation is occurring in existing areas, additional costs are incurred because of increased pavement reconstruction and the requirement to maintain customer supply during construction.

In some communities, this type of analysis is done in conjunction with the sewage and transportation departments to ensure efficient use of resources if all three infrastructure components need to be upgraded in the same year.

Existing neighborhoods should also be periodically assessed to determine whether the utility is providing the necessary fire flows. Over time, the type of development in a neighborhood changes as single-family homes are replaced with multifamily buildings. The carrying capacity of a pipe may degrade with age, depending on the corrosive nature of the water and soils. These two factors sometimes result in a water system that does not deliver the appropriate fire flows for the development. A detailed model of an existing neighborhood is used to identify the most cost-effective solution to upgrade the firefighting capabilities.

Outage Planning

If a community maintains a calibrated model of their transmission and distribution system, the model is used to assess the impact of an upcoming outage on the system. The modeler should assess three main issues: reservoir volumes, minimum system pressures, and firefighting capabilities.

Reservoir volumes are reduced either because of a reservoir shutdown for cleaning or maintenance or by the shutdown of a transmission main allowing the reservoir to be filled or to supply water to a neighborhood. In these cases, the duration of the shutdown is assessed against production capabilities and the network analyzed to ensure the needed water is moved to the other locations to maintain customer supply.

The minimum pressures in the water distribution network are analyzed during the planned shutdown. If there is a major change and the shutdown is expected to cover an extended period, residents in the affected area should be notified in advance. The utility call center is also be notified so qualified personnel can respond effectively to any customer complaints caused by the reduced level of service.

Finally, the firefighting capabilities in the immediate area of the shutdown are assessed. Contingency plans are developed with the local fire department to identify how to respond to a fire in the affected area during the shutdown.

Energy Optimization

The potential energy savings from energy optimization can be as much as 10 or 20 percent of the total energy cost. For this reason, evaluating energy requirements and developing a power-management program are key steps in the development of distribution system improvements.

Some of the items to consider in developing an energy optimization plan are as follows:

- power consumption by system components
- alternate system operations to reduce energy consumption
- peak shaving through standby power
- construction of additional pipes, pumps, or storage to reduce overall pumping requirements

The Energy and Water Quality Management projects sponsored by the American Water Works Association Research Foundation and the Electric Power Research Institute are good resources to assist the modeler in identifying opportunities for energy optimization.

Operator Training and Contingency Planning

Operator training is an excellent application of distribution system modeling within a utility. The models are used to train new staff on the unique operational characteristics of the water distribution system. The models are also useful in simulating different system failures and identifying the most effective response to reduce the impact on customers.

The operators are also the best resource for determining if the model accurately represents flow and pressure measurements based on their SCADA system. This allows the modeler and the operator to work together to identify system deficiencies or problems in the model assumptions.

CONTINUING USE OF THE MODEL

A utility that invests the time, effort, and money in performing a computer-assisted analysis of its water distribution system should consider maintaining the model for continued use. A model is a corporate asset and should be treated as such. This requires the support of a modeling team that ensures that the model remains current and sufficiently calibrated for its intended use. The integrity of the base model is paramount. One member of the team should be responsible for changing the base model and maintaining backup copies. Documentation of the model is critical to ensure that any new users are aware of the limitations of the simulation. Periodic checks against actual operating conditions help identify any changes in the distribution system that affect the accuracy of the model. If internal resources are not sufficient, the utility should consider contracting all or part of the model activities to outside consultants. It is essential that the modeler be properly trained in hydraulic analysis and computer modeling.

REFERENCES

Electric Power Research Institute (EPRI). 1999. *A Total Energy & Water Quality Management System*. EPRI, Palo Alto, Calif.

Cesario, A.L. 1995. *Modeling, Analysis and Design of Water Distribution Systems*. AWWA, Denver, Colo.

Chase, Don. 1999. Calibration Guidelines for Water Distribution System Modeling. In *Proceedings of the Information Management Technology Conference*. AWWA, Denver, Colo.

Cruickshank, J.R. and S.J. Long. 1992. Calibrating Computer Models of Distribution Systems. In *1992 Computer Conference Proceedings*. AWWA, Denver, Colo.

AWWA. 1998. Manual M31—*Distribution System Requirements for Fire Protection*. AWWA, Denver, Colo.

National Fire Protection Association (NFPA). 1985. NFPA Standard 291-1985, Recommended Practice for *Fire Flow Testing and Marking of Hydrants*. NFPA, Quincy, Mass.

Walski, T.M. 1995. Standards for Model Calibration. In *Computer Conference Proceedings*. AWWA, Denver, Colo.

Walski, T.M., and S.J. O'Farrell. 1994. Head Loss Testing in Transmission Mains. *Jour. AWWA*, 86:7:62–65.

Fire Underwriters Survey. 1981. *Water Supply for Public Fire Protection—A Guide to Recommended Practice*. Insurance Bureau of Canada, Toronto, Ont.

This page intentionally blank.

AWWA MANUAL M32

Chapter 5

Extended-Period Simulation

INTRODUCTION

Extended-period simulation is a way of modeling a distribution system where a series of steady-state simulations at specified intervals are performed over a time period to model the way a system changes in response to changing demands and operational conditions. Extended-period simulations greatly improve water system planning and operation at a relatively low cost. Extended-period simulation distribution system models are also used to refine water supply and distribution system improvements. Typically, analyses are simulated over several hours or days, such as a 24-hr period, during average and maximum demand days. These 24-hr period simulations are analyzed at user-selected time intervals, from several minutes to several hours.

In addition to steady-state analysis performed on water supply and distribution systems, an extended-period simulation can further analyze both existing and future water systems to test system-operating procedures. While a steady-state analysis is a useful tool to size pipelines and supply facilities, an extended-period simulation analysis provides a great deal more information about system operating characteristics and how the water system responds to changing demands or emergency situations. This includes information such as proper booster pump sizing, determining fluctuating reservoir levels, and more efficient system control valve settings.

Under continually changing consumption rates that water systems experience, storage facilities within a water system permit more uniform pumping rates over time, and hence, more efficient system operation. Reservoirs also provide reserves for firefighting and other water supply emergencies. For simple systems containing only one floating storage facility, water level fluctuations are easily calculated from incremental net changes in system inflow and outflow. However, in complex systems, reservoir level prediction and booster pump interaction becomes more complicated because of hourly variations and pressure zone demands on each reservoir.

For such complicated water systems, extended-period simulations are extremely useful in determining how a system behaves under changing demands, how a stressed system reacts in emergencies, or in determining the best locations and sizes of future storage and pumping facilities. For each reservoir, the volume used as floating storage, the minimum emergency reserve, and the speed of refilling are easily calculated. An extended-period simulation uses a steady-state analysis subroutine to determine pressures and flows in the distribution system and a subroutine performing a time integration of reservoir node flow in or out of the system to track reservoir volume changes.

When modeling individual pressure zones separately, it is impossible to accurately predict how boundary conditions of one zone impacts an adjacent zone under varying supply and demand conditions. Experience shows that entire water systems, with multiple pressure zones, must be modeled together in order to accurately simulate real water system operation when pressure zones are interconnected. For these complex interconnected water system pressure zones, extended-period simulations model how an entire water system interacts dynamically via varying reservoir water levels, booster pumps turning on and off, and pressure and flow control valve operation between pressure zones.

INPUT DATA

All of the data in a steady-state simulation are also used in an extended-period simulation. In addition to the data used in a steady-state simulation, additional data is required to describe how demands, pumps, valves, reservoir levels, and water quality parameters change over time. This section describes the additional information useful for an extended-period model.

Verifying System Operation

Results of steady-state computer analyses on an existing system are first verified to ensure that they agree closely with actual system operations. Only then should extended-period simulations be conducted. Booster pump data in an extended-period computer run is input to turn pumps on and off in similar sequence as in the real system. Reservoir water levels during an extended-period analysis should agree with historical field observations of water level fluctuations. Regulating valves and general distribution system pressures contained in computer output are also similar to conditions experienced in the field.

Establishing the Extended-Period Simulation Time Interval

The extended-period simulation time interval is set based on two factors: the interval necessary to achieve the objectives of the simulation, and the interval that is appropriate for equipment controls to function properly. A reservoir storage analysis, for example, might run for 24 hr and have a computation time interval of 30 min or 1 hr. A water quality simulation could run for several days with computation time intervals that are accurate. Changes in equipment controls are often done by patterns, curves, or profiles that describe the changes that are to take place over time. The computational time interval should not be so long that significant changes in the patterns are missed. On complex models, the time interval also should not be so short that the time to complete an extended-period simulation run is unnecessarily long.

Booster Stations and Wells

As with steady-state simulations, extended-period simulation data includes actual pump performance curves used in the input data. Booster stations having multiple pumps operating on system pressure also contain control information to allow the computer program to determine when pumps turn on or off. A simple method of modeling multiple pumps is to model each pump individually and establish settings to simulate actual pump settings in the real system. As system pressure or reservoir level drops, additional pumps "turn on" in the model. Whenever a modeled pump turns on or off, basic system boundary conditions change. Therefore, an extended-period calculation and output are generated at that specific time interval. Care should be taken to not set pump pressure setting too close, to avoid several short-duration time interval calculations. Also, modelers should check that pump starts and stops do not adversely affect other modeled components, such as other nearby system booster pumps and control valves and that system pressure data accurately reflect real system reactions.

Tanks and Reservoirs

To model reservoir water level fluctuations, specific tank input data are required. These data include maximum and minimum hydraulic grade line elevations, tank diameter, and external sources of inflow or outflow, such as any wells and booster stations that are connected directly to a reservoir. For noncircular reservoirs, a depth-to-volume relationship is calculated based on reservoir capacity and water surface height.

By calculating flow rates into and out of reservoirs at each time interval during a simulation, a volumetric change in storage is computed, and a new water surface elevation calculated. Simulations begin at any time interval. However, it is often useful to begin simulations at midnight or some other low-demand period and allow the water system to cycle through an entire 24-hr day, or other "typical" operating cycle. Reservoirs are typically at or near full late at night. Normal reservoir water level fluctuations during daily demand periods can range from 5 to 15 ft (0.6 m to 4.6 m) of water depth. Starting reservoir water levels are set based on actual reservoir operating records for typical average and maximum day demands. In the absence of actual records, reservoirs are assumed to be at or near full at midnight.

Control Valves

Proper control valve operation is critical to performing accurate extended-period simulations. Typical control valves include pressure-reducing or sustaining valves, flow control valves, and altitude valves. Proper pressure or flow settings are important if these elements are to accurately represent the real system. As with pumps, whenever these elements open, close or modulate, extended-period simulation time interval calculation occurs. The modeler should carefully check how the other system elements respond to these changes. For example, an altitude valve closing when a reservoir reaches its maximum water level results in instantaneous system pressure increases of up to 10 psi (69 kPa) or more, thus causing system pumps or control valves to change operation. While this may be a normal response in the real system, such model results should always be verified by actual operating data or discussions with operations personnel.

Other Supply Sources

Connections to regional water supply systems, major transmission pipelines, and interconnections to other water distribution systems are modeled by giving each node corresponding to these sources a constant flow rate into the system. Flow rates should be chosen that represent typical average flow rates for the demand condition specified.

Emergency System Operations

An extremely valuable benefit of extended-period simulations is in evaluating system response under emergency conditions, such as simulated major fires during a maximum-day demand condition or simulated power outages. These "what if" analyses provide operations personnel important information on how water system facilities respond when stressed. This enables advanced planning for better response in emergency situations, if and when, they occur.

Major industrial areas typically require very large fire flows. To simulate system response under worst-case conditions, these large fire flows are set in early morning hours, coinciding with high morning demands. Similar to a simulated fire, system power outages are also assumed during critical demand periods to determine system response. Under such circumstances, wells and pumping stations without standby power sources would be out of service. Therefore, the only water available comes from supplies having emergency power generators, system storage, or emergency interconnections with other systems.

During emergency simulations, system response should be analyzed for areas having below standard pressures (less than 20 psi [138 kPa]), negative pressures, and low, or empty reservoir levels. For multizone systems, interzone water transfers should be analyzed to ensure that the whole water system responds to emergency conditions. These simulations are useful tools in changing operating procedures to improve system performance.

EXTENDED-PERIOD SIMULATION SETUP

Objective

Steady-state water system modeling was originally designed to describe system performance under a variety of different demand conditions (e.g., maximum system conditions, such as maximum-day demand plus fire load, peak hour demand, or conditions in which system demand is at its minimum while water production and water transfer to storage is at its maximum). It is important to identify and remedy major system performance deficiencies. System components requiring calibration include piping configuration, size, demands, and coefficients of friction to accurately describe friction losses associated with transporting vast quantities of water through the system. The steady-state evaluation is essentially a snapshot of the system under a defined demand situation. Extended-period simulation is an extension of the steady-state analysis in that it performs a series of evaluations that verify system performance. Extended-period simulation calibration involves undergoing a certain series of conditions to emulate a more fluid look at the system through the system demand fluctuations that occur over a 24-hr, or even greater, time period. System calibration requirements for this evaluation include the steady-state calibration requirements but extend to additional information mandatory for accurate system evaluations. This section discusses some of these additional requirements required for extended-period simulation.

System Operational Data

System operational constraints, either imposed by physical conditions or by policy designs, must be identified so that their impact on the system operation and performance can be evaluated. This system operation, in the form of information requirements, is described in the following sections.

Overall Operational Philosophy

Utilities establish an acceptable level of service to satisfy customer demands. Most utilities have a goal of providing continuous delivery of high-quality drinking water to all customers. This goal assumes that adequate pumping, storage, and redundant critical system components are available so it can be achieved even during emergency operations (e.g., firefighting or natural disasters). Regular water outages or other lengthy interruptions in water service are not tolerated in most US communities.

Peak Day Philosophy

Often, a water system is designed to sustain continuous peak (or maximum) day demands, requiring that the supply over a 24-hr period equals the total system demand over that period. While available water storage is capable of supplying instantaneous demands over peak day supply and storing supply in excess of low-usage periods, this philosophy assures that the system meets the system demand under any conditions and requires that the supply is equal to the peak day demand requirement but minimizes the storage requirements.

Peak Week Philosophy

If the supply is constrained, or the possibility of a peak day event experienced over a series of consequent days is not great, the system should be designed for a longer period of recovery. Again, the supply over a 7-day or greater period equals the total system demand over that period while the storage is capable of supplying instantaneous demands over the available supply and storing supply in excess of low-usage periods. This philosophy allows a smaller percentage of supply relative to peak day demand but requires additional storage.

Time-of-Day Philosophy

Power utilities may opt to make more attractive rates available to the water utility if the demand for power occurs during the power utility's off-peak hours. This Time-of-Day rate brings significant energy cost savings but possibly at a significant capital cost to the water utility system.

Reservoir or Service Storage Data

Static distribution system models are concerned with water surface elevation at a particular point in time. One of the most important pieces of information required for extended-period simulations is the amount of service storage available and the characteristics of that storage. Vital information developed for the model includes

- Overflow elevation—the maximum water surface elevation
- Storage/unit of height—the amount of water contained per unit of height
- Effective height—the difference in elevation between the maximum and minimum water surface elevation
- Reservoir physical attributes—uniformly circular, rectangular, irregular at bottom or top

Pump Station Information

Pump station input requirements are also significantly more detailed in the extended-period model than the steady-state model. Accurate individual pump curves are essential for an accurate extended-period model, because minor fluctuations in suction pressure or service reservoir levels greatly affect model predictions. The actual station operating parameters, or control set points and location of the controlling point in the system (reservoir or system junction point), must be input into the model, depending again on the accuracy required. The pertinent information for the station includes the following:

- Certified pump curves for each pumping unit within the station pump; controlling point location
- Set points
- Initial status (on/off) for each pump

Well Data

The well pump level and pump curve are generally input into the extended-period simulation model. The extended-period simulation model also requires the location of the controlling point in the system (reservoir or system junction point), the actual on/off set points, and initial condition be input to allow the well to function in the model as it would in the real world.

SCADA Information

Extended-period simulations are designed to mimic water system operation with respect to system demand over time. The more precise and detailed the information, the greater the model accuracy and greater the potential rewards of the evaluation. Essential to accurate system analysis is obtaining information that is characterized as two basic components. The first is an accurate description of system demand fluctuations over some finite time, optimally hourly or less. The second is a detailed description of the operation parameter changes (pump on/off) and reaction of the water system to those system demand fluctuations (storage volume changes) over the same finite time. System status information obtained every 24 hr is essential in describing system peak day events for a water system and useful in static model evaluations, but insufficient in developing an extended-period simulation model. Central in obtaining system operation parameter changes and reaction to the water system is a SCADA system that monitors operation parameters as they change and stores information in an easily retrievable form for use in the model. This information, broken into finite time intervals, such as hourly changes at a minimum, is the standard to measure system performance versus model predictions. The basic SCADA information required for extended-period simulations is listed below:

- Reservoir fluctuations over time
- Pump run times, flow variations
- System pressure fluctuations over time
- System control valve status, pressures and flow, if, any

Diurnal Curve Characterization

Derivation of diurnal curves (see chapter 3) for major customer types is important when performing steady-state analyses. The accuracy of an extended-period simulation model is highly dependent on the quality of the demands and diurnal patterns assigned to the model. To create an accurate extended-period simulation model, it is no longer sufficient to broadly characterize zones or areas based on assumed demands. Specific water demands and diurnal patterns are needed to ensure the accuracy of the model. The specific water-use characteristics of the residents in that neighborhood are described in order to accurately predict water use over time. Homes with large landscaped lots demonstrate outside watering characteristics far different than homes with water-efficient landscapes. Subtle differences in water use translate into inaccuracies in the model if not included in the diurnal curve for each customer classification.

The dichotomy of determining or describing water system demands lies in the different areas in which water use data is collected or stored. Virtually all larger water systems have personnel whose specific duty is to read on a regular basis water meters for all customer types. This data may or may not be stored in a database easily retrievable for use in analyzing water consumption, but it contains only the information for the time frame when the meter is read, usually monthly or sometimes quarterly. The sole purpose of this duty is to charge customers their share of costs in producing the water based on their specific usage. Conversely, many utilities have sophisticated SCADA systems with the ability to acquire and retain system information. This information, often stored in at least 30-min intervals, is nothing more than a description of how the water system reacts to system demand. There is no direct correlation between the two water system usage data collection systems.

Few utilities have the personnel or the budget to install a significant number of water meters sophisticated enough to monitor water usage for the broad type and number of customer types required to characterize water demand over small increments of time concurrent with SCADA information. However, equipment can be purchased to attach to existing meters that collects diurnal demand information. If necessary, meters are manually inspected over a 24-hr period to gather information. Developing a good set of demand data significantly improves the accuracy, and therefore the value, of a hydraulic model.

Aside from taking an exhaustive series of actual field measurements representative of the various customer types throughout the entire service area, the only remaining option involves using data available from billing records and SCADA information; the meter usage data are classified by customer type or geographical area and assigned a diurnal curve based on flow information from SCADA.

Model Calibration Process

The model calibration process is one in which model runs are compared with recorded data. Model data that directly impacts the results are described as system configuration, system condition, system reactions to demand, and actual demand itself. Two of these components, namely system configuration and condition, should have been calibrated during the steady-state water model calibration process.

System configuration is a description of the system components, including pressure zone configuration, network pipe size, system architecture, valving, pressure regulating valve location and set points, and pipe connectivity. System condition involves the amount of energy required to transport water through the system. Both of these components should have been determined at least to some extent during the static water modeling process.

The other major components are the water system demand over time and the reaction of the water system, contained in the SCADA information, to those demands. These two components are dependent on each other.

Again, extended-period simulation models are only as accurate as the model input data. Many software models on the market contain some ability to input facility control set point data. Others actually allow sufficient facility set point information, in terms of control set point algorithms or control logic, that the model reacts to system loading exactly as the water system should theoretically react. These new tools allow the modeler the ability to "mimic" the SCADA information.

If all aspects of the physical water system are accurately described, the only remaining information required is an accurate description of system demand over time. Current water models generally allow a multiple number of diurnal curves to describe water use patterns for the various user types. Extended-period simulation model calibration essentially involves refining the water-use diurnal curve characterization for the various user types and knowing the base demand for those user types on which to apply the diurnal curves.

Few water system service areas consist of a homogeneous customer type. However, it is probable that most contain a few areas that are characterized as homogeneous and about which the actual system demand is described. For example, if a residential area can be found whose water use is segregated from the remaining system and whose water-use characteristics are described using SCADA records or field measurements using temporary flowmeters, a diurnal curve for this residential type can be described. A similar residential area, but combined with some multifamily or commercial use, is evaluated using the previously determined residential diurnal to filter out the residential demand, thereby describing the multifamily or commercial diurnal.

Once diurnal curves are derived for the various customer types, it is necessary to apply base demands, either as individual users or as a block or blocks of similar individual users, to the developed diurnal curves and populating the node influence areas in the model. Once this is accomplished the calibration effort consists of a series of comparisons between model-derived estimates of major facility changes over time versus actual facility changes (reservoir levels and pump run times) described in the SCADA information.

Example Uses of Extended-Period Simulation Analysis

The extended-period simulation analysis technique, performed after the model is calibrated, is tailored based on the objectives of the actual project. Modeling criteria must first be described on which the actual analysis technique is determined. Several of the more typical analysis objectives are discussed in some detail in the following. The results of a real water system analysis are provided to illustrate the impact these analyses have in a real-world situation. The particular system involves interaction of two reservoirs, each with a well-defined service area and containing two very distinct diurnal demand curve types. The system demand in this example contains two separate components, a residential demand and a golf course demand. Figure 5-1 represents a peak-day diurnal curve for the entire residential population, based on the residential diurnal curve established for the residential population, adjusted to the actual residential population served.

Figure 5-2 is a graph that represents the golf course load over the same time period, based on its diurnal curve and the actual instantaneous demand.

EXTENDED-PERIOD SIMULATION   87

Graph 1

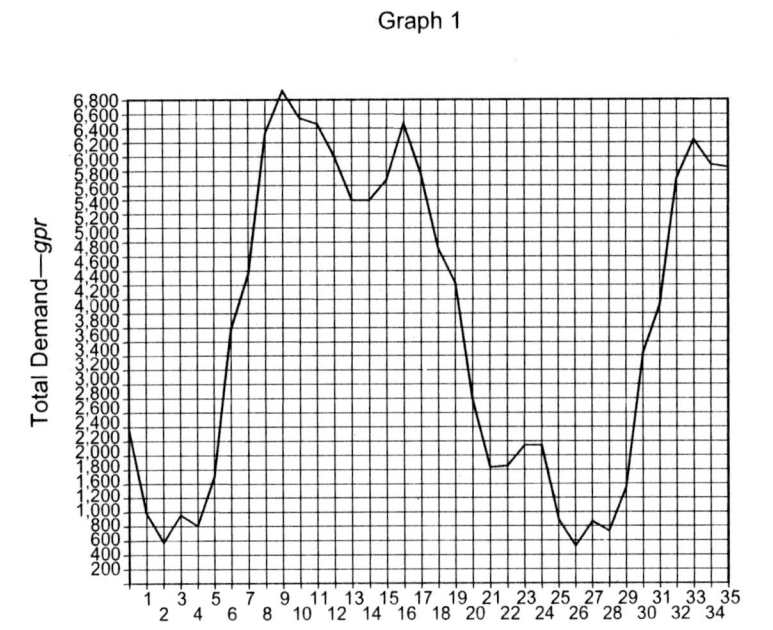

Figure 5-1   Residential system demand vs. time

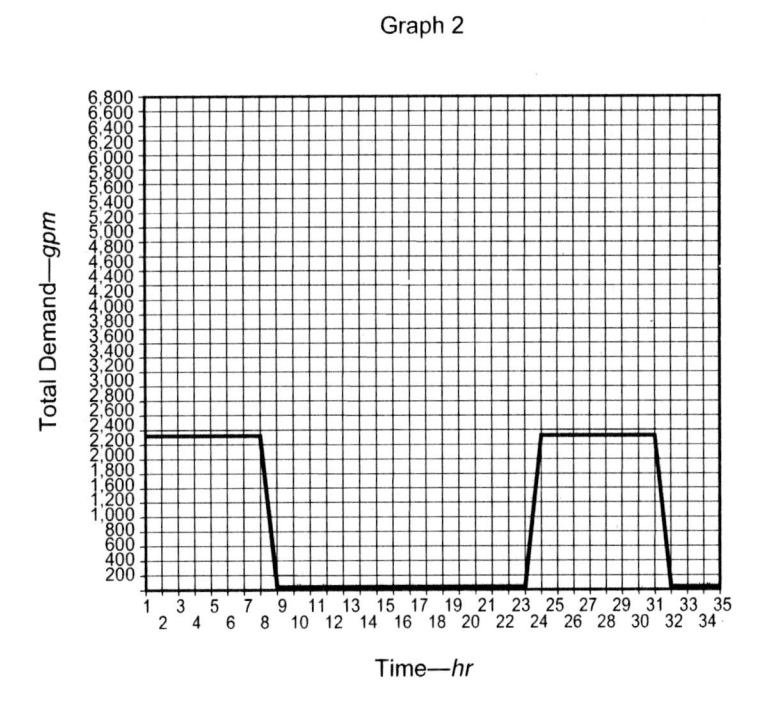

Graph 2

Figure 5-2   Golf course system demand vs. time

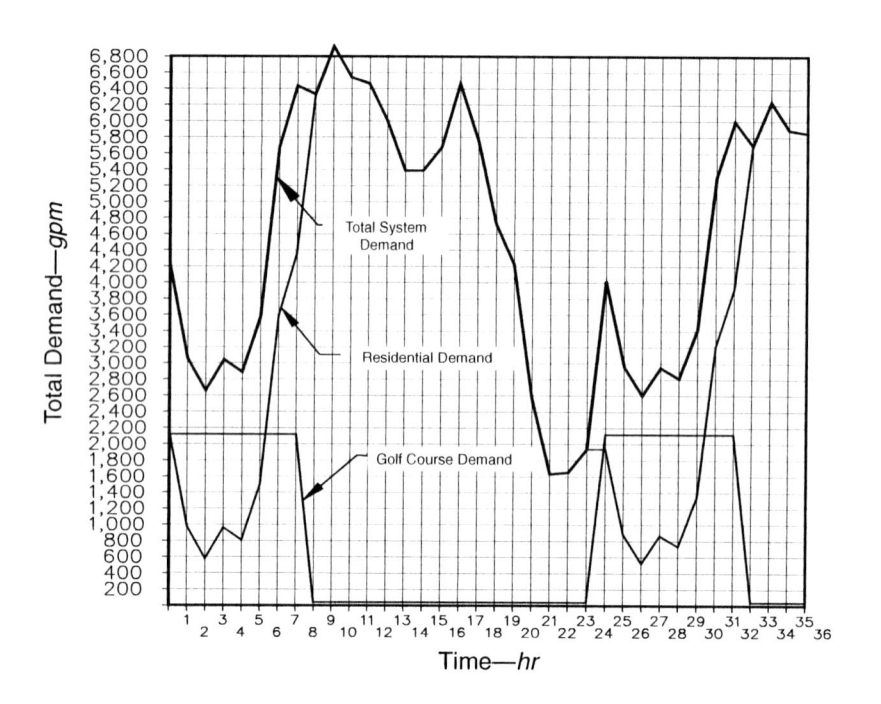

88   COMPUTER MODELING OF WATER DISTRIBUTION SYSTEMS

Figure 5-3   Total system demand vs. time

In an extended-period simulation evaluation these two loads are added, as in Figure 5-3, to form the total system diurnal curve for the period.

The two reservoirs each have a well supply on-site, plus a single well within the service area. The system physical parameters are as in Table 5-1:

Table 5-1   System physical parameters for extended-period simulation analysis

Reservoir No. 1		Reservoir No. 2	
Overflow elevation	5,485.00'	Overflow elevation	5,485.00'
Bottom elevation	5,453.00'	Bottom elevation	5,450.63'
Effective storage	2.0 MG	Effective storage	3.08 MG
Storage/ft	62,500 gal	Storage/ft	86,613 gal
Well No. 1	1,800 gpm @ 563' of lift		
Well No. 2	1,450 gpm @ 563' of lift		
Well No. 3	2,075 gpm @ 563' of lift		
Fire Protection requirements:	5,000 gpm for 5 hr or 1.5 MG		

Storage vs. Production

The most basic extended-period simulation evaluation involves determination of storage versus production, again with the premise that the water system must return to an initial condition within a prescribed time frame. For the example of storage versus production, as in Figure 5-4, it is assumed that status quo must be maintained in a 24-hr period. That means the constant production over a 24-hr period equals the total system demand over that same time frame, while the storage must either make up demand over production or store production over demand. Storage and production must combine to provide service while returning the system to status quo at the end of the 24-hour period. Because the loads are not symmetric with respect to storage of production, a combination of storage volumes plus water system piping must be sized appropriately.

The example utility contained originally only reservoirs 1 and 2 and wells 1 and 2. An analysis was performed using the 24-hr peak day diurnal curve followed by an additional 12 hr of 90 percent peak day to determine the adequacy of storage and production.

As is noted in Case 1 data (Figure 5-4), the water system could not completely recover for the second day, resulting in a situation in which the reservoir came close to storage set aside for fire protection.

This situation is alleviated by introduction of additional production, as shown in Figure 5-5. An identical evaluation was prepared using Well 3 as a supplemental well to the system. Case 2 (Figure 5-5) illustrates the effect on the system of Well 3, set to operate when the reservoirs approach a water level 10 ft below overflow. The reservoir water elevations returned to a 24-hr full situation due to the additional supply from Well 3.

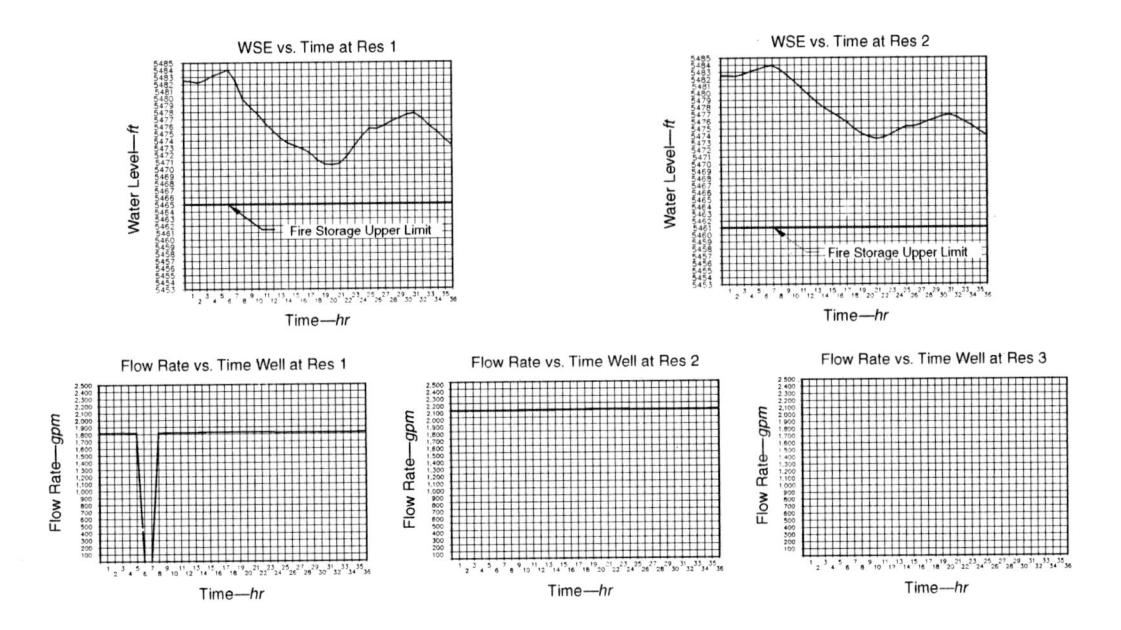

Figure 5-4 Example of storage vs. production, Case 1

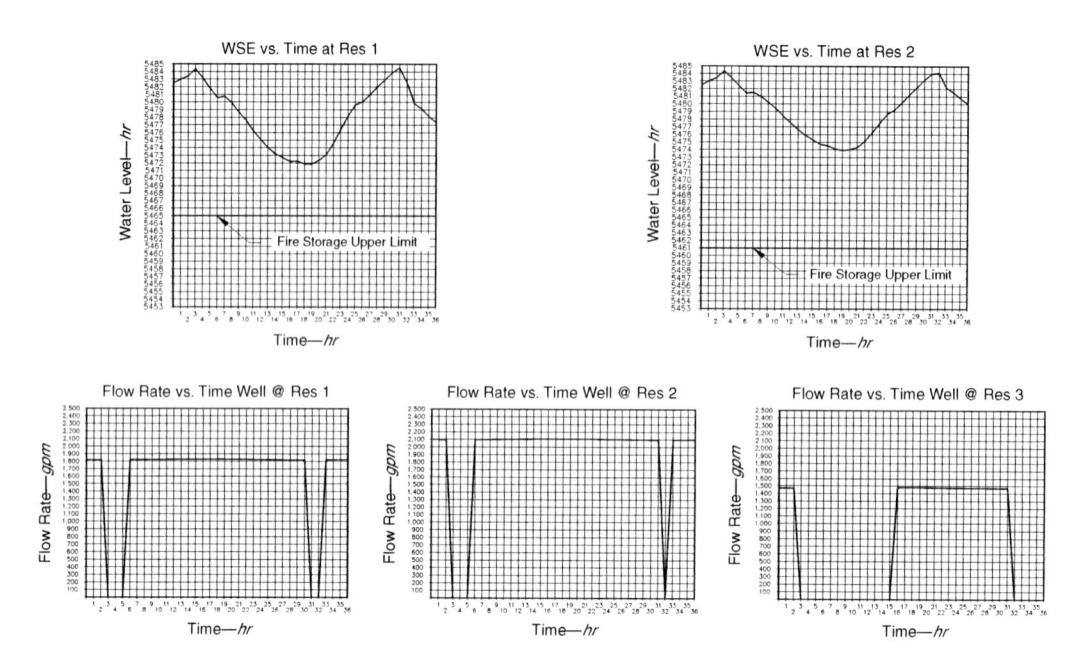

Figure 5-5 Example of storage vs. production, Case 2

Vulnerability Analysis

Another key evaluation performed using extended-period simulation is reliability or vulnerability analysis. This analysis is used to identify critical weak points within the system, assess the extent of those weak points and effect on water service, and determine system modifications that provide some relief from the problem. This relief might result in some redundant production, storage, or a combination. For example, the loss of a well at the Reservoir 1 site results in reservoir elevations as shown in Figure 5-6, assuming all other assumptions constant. The system that was balanced previously is now unbalanced to the point that Reservoir 1 level compromises fire storage and, if a second peak day event occurred, would drain.

ENERGY OPTIMIZATION

Energy optimization is very important for efficient water system operation because of the rising cost of energy. Energy optimization involves a delicate balance between supply, pumping, and storage based on cost differentials between various energy sources, either by type or by time of service. Several types of energy optimization techniques are described in the following sections to illustrate the effects of time-of-day rates and blending energy types on water service or system capital requirements.

Time-of-Day Electric Rates

Electricity generation, unlike water supply, must meet or exceed instantaneous electric demand. This fact results in significant capital expenditures for peak electric generation equipment that are idle the vast majority of time. Some electric generation companies have opted to smooth the system load over nonpeak times, targeting large users by rewarding them for off-peak electric demands and penalizing them for energy demand that coincides with the system peak electric demand. With

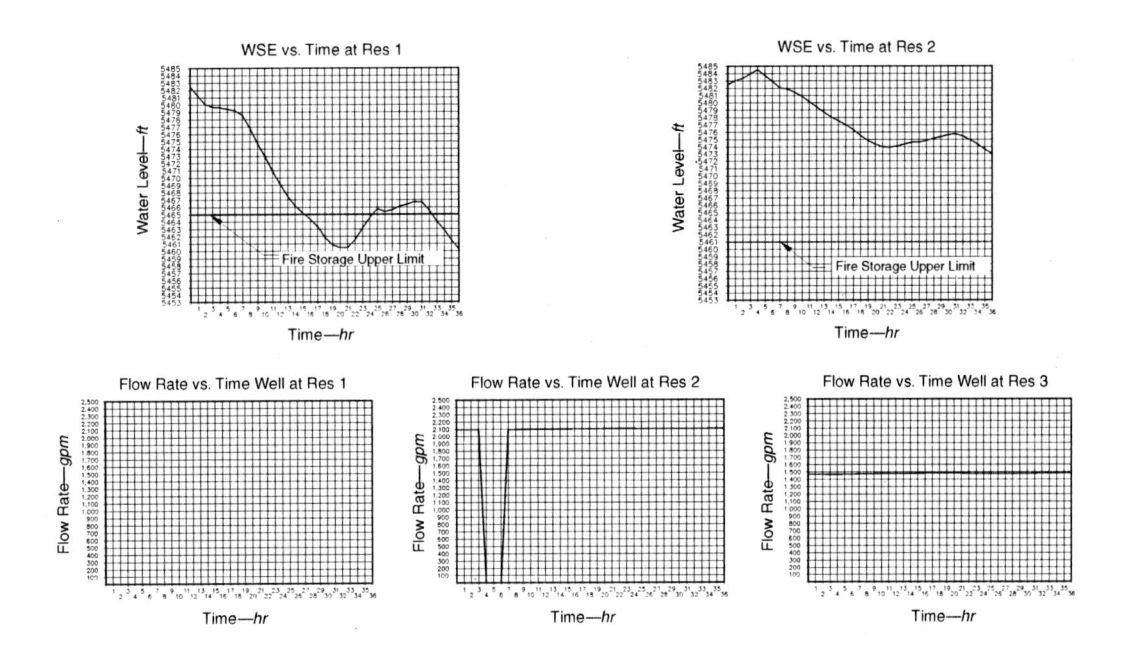

Figure 5-6 Example of storage vs. production vulnerability analysis, Case 3

few exceptions, water utilities comprise a community's larger—if not largest—electric user, and water system demand peaks usually mirror electric demand peaks; in fact, a significant percentage of water system demand occurs during a 12-hr peak electric demand time frame. Figure 5-7, illustrates the same example using a 12-hr pumping curtailment from 8:00 a.m. to 8:00 p.m. Reservoir 1 actually drained completely during the 36-hr period while Reservoir 2 certainly is into fire storage. The result is a lack of service to some and insufficient fire protection to all.

Water utilities utilizing only electric power have options available to them to curtail on-peak electric consumption. The first involves a balance of additional storage and supply to make up the deficit. The second involves an analysis to determine the actual number and extent of system peaks expected to occur and balance the cost of the additional infrastructure versus the additional energy costs incurred in violating the on-peak pumping for those periods of the year.

Driver Types/Blend

One way to use electric time-of-day rates is to supplement electric power needs by using other drivers. A natural gas engine connected to a well pump can provide water during on-peak electric periods and supplement electric drivers off-peak. In addition, peak water demand occurs during the growing season at a time when natural gas energy requirements for heat are traditionally at low levels. Figure 5-8 illustrates the impact of changing Well 3 to a natural gas driver. Clearly, the additional production of Well 3 during the on-peak electric period is not sufficient to offset the storage losses caused by the other wells being off, but there is a significant improvement over the reservoir levels illustrated in Figure 5-7.

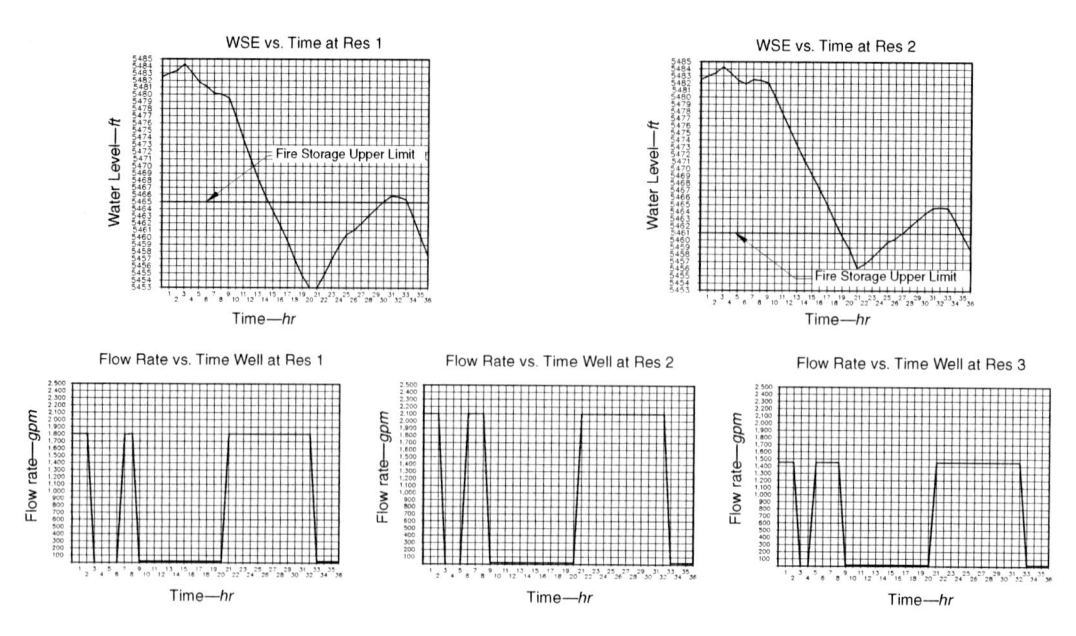

92 COMPUTER MODELING OF WATER DISTRIBUTION SYSTEMS

Figure 5-7 Example of storage vs. production with pumping curtailment, Case 4

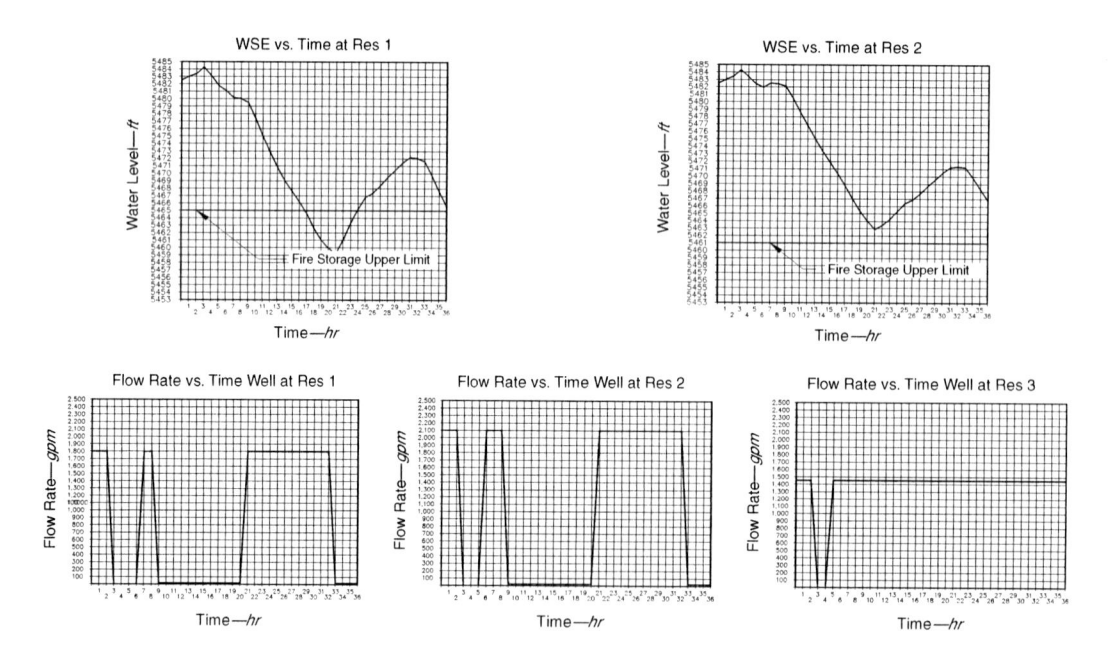

Figure 5-8 Example of storage vs. production with supplemental power, Case 5

CASE STUDY CITY OF FULLERTON, CALIF.

In a study conducted for the city of Fullerton, Calif., a time-dependent computer model identified necessary new facilities and operational improvements to the water supply system that saved the city considerable operating costs. An extended-period simulation enabled the city to identify ways to maximize low-cost well water, and reduce the amount of pumping (and energy required) throughout their water system.

The city of Fullerton receives its water supply from wells in the local Orange County Groundwater Basin and from water imported by the regional water supplier, Metropolitan Water District of Southern California (MWD). City wells pump to forebays or reservoirs, and booster pumping stations pump the water into the system. A few wells pump directly into the system. MWD delivers imported, treated water from the Colorado River and California State Water Project. MWD's transmission system provides metered connections to the city's distribution system. These connections are controlled by pressure-regulating and flow-control facilities.

The city has a complex distribution system composed of 7 MWD connections, 10 storage facilities, 12 wells, 11 booster pumping stations, and 28 pressure-regulating stations. The service area is divided into 13 pressure zones. The zones are broken into four main and nine smaller, subpressure zones. Storage reservoirs and booster pumping stations equalize flows between and maintain system pressures within each zone. All zones are interconnected through pressure-regulating stations that maintain minimum zone pressures.

The basic operational scheme involves pumping well water into lower-elevation pressure zones and pumping this water into successively higher-pressure zones. Most MWD water enters the distribution system through higher pressure zones. Imported water is allowed to flow into lower pressure zones, based on system demands, through pressure-regulating stations. In this manner, well water is pumped upward and imported MWD water generally flows downward through the system.

The time-dependent model included all major water system facilities: transmission and distribution pipelines typically 8-in. and larger, storage reservoirs, booster pumping stations, wells, MWD connections, and the more actively used pressure-regulating stations. Physical features of these water system facilities were defined in mathematical form and entered into the computer model. Numerous time-dependent (24-hr time period) computer simulations were performed under various demand conditions to evaluate the behavior of the network.

Existing Inefficiencies

Computer simulations pinpointed a number of potential operating improvements within the existing system. For example, a reservoir in Zone 1, the lowest pressure zone, maintained an unusually high water level throughout maximum demand conditions and therefore, never fully contributed to the system. Investigation revealed that the pressure gradient setting at a nearby pressure-regulating station was set too close to the reservoir's high water level to allow adequate water circulation. The reservoir was continually filled from this higher pressure zone. Lowering the pressure setting allowed much more flow from the reservoir during high demands.

In another example, the computer analysis showed that water pumped from another Zone 1 reservoir into Zone 3, a higher zone, was recirculating through a nearby pressure-regulating station to supply an intermediate pressure zone, Zone 2. More water should have been pumped from the reservoir directly to Zone 2 where it was needed, rather than to Zone 3. Once booster pumps were adjusted, the recirculation between pressure zones and associated wasted pump energy was eliminated.

Operational Improvements

Having identified deficiencies in the existing system, future demand conditions were analyzed. Once future water system improvements were selected, time-dependent analysis was used to determine the most efficient water supply system operation under future conditions. Basic study objectives were to (1) minimize interzone water transfer and resultant energy lost through pressure-regulating stations; (2) maximize lower-cost groundwater production; and (3) determine reservoir storage capacity needed to meet regulatory, emergency, and fire flow requirements of the future system.

Pressure-Regulating Stations

The computer model showed that interzone water transfers could be greatly reduced without adversely affecting system pressure or supply by lowering downstream pressure gradient settings at pressure-regulating stations and adding to or modifying supply facilities. With these changes, an analysis of future demand conditions showed much less water transfer during days of maximum demand. Overall, interzone water transfers were reduced approximately 80 percent compared with existing system operation.

Booster Stations and Wells

A major objective of the extended-period simulation was to identify ways to meet most of the city's water demand from groundwater sources instead of imported MWD water. To explore conditions for optimum groundwater production, new wells and additional booster pumping capacity at existing wells were added to the model and the results were studied. The computer model showed that two new wells and an increase in well booster pumping capacity to match production capacity could produce enough groundwater to meet most of the systems' winter and nighttime demand. Also, an increased amount of groundwater could be mixed with MWD water to meet summer demand. A new booster station to pump water from Zone 1 to Zone 3 was also recommended to increase the amount of well water pumped to higher pressure zones and help solve a fire flow deficiency.

Reservoirs

To determine reservoir storage capacity requirements, future needs of the city were studied, as well as individual needs of each pressure zone. Generally, citywide storage capacity was adequate to meet future needs. However, some individual pressure zones required additional storage capacity because of limitations in interzone water transfers. Here, time-dependent computer simulations showed that pressure Zone 2 required additional emergency storage to meet ultimate demands. Also, although Zone 3 had adequate storage capacity, there were physical constraints in the distribution system piping that, in effect, isolated a large part of the zone without adequate emergency storage capacity. Therefore, more emergency storage was proposed for that area.

REFERENCES

Bhave, P.R. 1991. *Analysis of Flow in Water Distribution Systems.* Technomics Publ., Lancaster, Pa.

Cross. H. 1936. *Analysis of Flow in Networks of Conduits or Conductors.* Univ. of Illinois, Urbana, Ill.

Eggener, C.L. and L. Polkowski. 1976. Network Modeling and the Impact of Modeling Assumptions. *Jour. AWWA* 68:4:189–196.

Gupta, R. and P. Bhave. 1996. Comparison of Methods for Predicting Deficient Network Performance. *Journal of Water Resources Planning and Management* 122:3:214.

Jeppson, T.W. 1976. *Analysis of Flow in Pipe Networks.* Ann Arbor Science Publishers, Ann Arbor, Mich.

Jordan., R.A, Priest, M., Jain, D.K. Jacobesen, L.B. 1999. Master Planning Utilizing H20NET Extended Period Simulation and Microsoft Access Generated Summary Reports. In *Proceedings of the Information Management and Technology Conference.* AWWA, Denver, Colo.

Shamir, U., and C.D.D. Howard. 1968. Water Distribution Systems Analysis. *Journal of the Hydraulics Division.* ASCE, 94:1:219.

Stone, K., Barbato, L.M., Koval, E.J. 2002. Colorado Springs Utilities: A Case Study for Extended Period Simulation. In *Proceedings of the Information Management and Technology Conference.* AWWA, Denver, Colo.

Todini, E., and S. Pilati. 1987. A Gradient Method for the Analysis of Pipe Networks. International Conference on Computer Applications for Water Supply and Distribution. Leicester, UK.

Walski, T.M., Gessler, J., and J.W. Sjostrom. 1990. *Water Distribution—Simulation and Sizing.* Lewis Publishers, Ann Arbor, Mich.

Wood, D.J. 1980. *Computer Analysis of Flow in Pipe Networks.* Univ. of Illinois, Urbana, Ill.

Wood, D.J. and C.O.A. Charles. 1972. Hydraulic Analysis Using Linear Theory. *Journal of the Hydraulics Division American Society of Civil Engineers.* ASCE, 98:7:1157–1170.

Wood, D.J. and A.G. Rayes. 1981. Reliability of Algorithms for Pipe Network Analysis. *Journal of the Hydraulics Division.* ASCE, 107:10:1247–1248.

WRc Plc. 1989. *Network Analysis—A Code of Practice.* Water Research Centre, Swindon, United Kingdom.

This page intentionally blank.

AWWA MANUAL M32

Chapter 6

Water Quality Modeling

INTRODUCTION

A primary goal of water distribution systems is to deliver potable water when and where it is needed. Ideally, there should be no change in the quality of water from the time it leaves the treatment plant until the time it is consumed. In reality, significant changes occur as water travels through a distribution system.

Under normal operating conditions these changes occur for several reasons. Chemical reactions that began at the treatment plant continue in the distribution system. A prime example is reactions involving residual disinfectant, such as chlorine. These reactions reduce the concentration of the disinfectant over time and also result in production of undesirable by-products, such as chloroform or taste- and odor-producing compounds. Another example is the formation of calcium carbonate or other coatings along pipe walls. Blending of waters from different sources in the distribution system produces changes in pH and chloramine speciation. Water also reacts with the walls of the pipe through which it flows. Corrosion and subsequent buildup of tuburcles and oxide coatings are typical examples. Pipe walls also support the growth of attached colonies of microorganisms or thin biofilms. When this growth is excessive, it often increases disinfectant demand, produces taste- and odor-causing compounds, and offers protection for opportunistic pathogenic organisms. One other cause of water quality degradation is accidental or intentional contamination.

Need for Water Quality Modeling

The spatial and temporal variation of water quality in a distribution system is often very complex because of both the hydraulic mixing patterns and the water quality transformations that occur within the network. The topological complexity of pipe networks in distribution systems and the ever-changing patterns of water usage rates over a day and over different seasons creates myriad pathways that water travels on its journey from treatment plant to consumer. Individual parcels of water lose their identity as they mix with other parcels of different residence time and quality throughout the system. Thus, the water quality reaching a tap may vary significantly over the course of a day or between seasons, and the water quality at two different taps, even if they are in close proximity, can display significantly different water quality because of the travel pathways and transformations between the treatment plant and the taps.

The complexity of flow paths and transformations makes it difficult, if not impossible, for a person experienced in a particular distribution system to determine travel times, blending, and subsequent water quality throughout the system. Mathematical models of the hydraulics and water quality of a system, after careful application, can provide a method for predicting the movement, transformations, and water quality in a network.

Uses of Water Quality Modeling

Water quality models are used to predict the spatial and temporal distribution of a variety of constituents within a distribution system. These constituents include the following:

1. The portion of water originating from a particular source
2. The age of water (i.e., how long since it left the treatment plant)
3. The concentration of a nonreactive tracer compound either added to or removed from the system (e.g., fluoride or sodium)
4. The concentration and loss rate of secondary disinfectant (e.g., chlorine or chloramines)
5. The concentration and growth rate of DBPs (e.g., trihalomethanes)

The ability to model the transport and outcome of these parameters helps system managers perform a variety of water quality studies. Examples include the following:

1. Calibrating and verifying hydraulic models of the system through the use of chemical tracers
2. Locating and sizing storage tanks and modifying system operation to reduce water age
3. Modifying system design and operation to provide a desired blend of waters from different sources
4. Finding the best combination of (a) pipe replacement, relining, and cleaning; (b) reduction in storage holding time; and (c) location and injection rate of booster stations to maintain desired disinfectant levels throughout the system
5. Assessing and minimizing the risk of consumer exposure to DBPs
6. Assessing system vulnerability to incidents of external contamination
7. Designing a cost-efficient routine monitoring program to identify water quality variations and potential problems
8. Identifying ways to reduce water age
9. Determining the degree of intermixing of multiple water sources
10. Identifying the most effective sampling locations in a distribution system
11. Identifying response times for various contamination scenarios
12. Determining the best location and dosage for rechlorination stations

Efforts to extend water quality models to include such features as predicting biofilm buildup and transport of suspended particles (such as microorganisms) are still in the research stage. At this point in time, water quality models are not used to directly assess such distribution system problems as minimizing biofilm growth, avoiding coliform outbreak occurrences, and avoiding red water problems.

Governing Principles of Water Quality Modeling

Just as hydraulic models conform to the conservation laws of fluid mass and energy, water quality models are based on conservation of constituent mass. These models represent the following phenomena occurring in a distribution system (Rossman et al. 1993).

Advective Transport of Mass within Pipes. A dissolved substance travels down the length of a pipe with the same average velocity as the carrier fluid while at the same time reacting (either growing or decaying) at a given rate. Longitudinal dispersion is usually not an important transport mechanism under most operating conditions. This means that the modeler can assume that there is no intermixing of mass between adjacent parcels of water traveling down a pipe.

Mixing of Mass at Pipe Junctions. All water quality models assume that at junctions receiving inflow from two or more pipes, the mixing of fluid is complete and instantaneous. Thus, the concentration of a substance in water leaving the junction is simply the flow-weighted sum of the concentrations in the inflowing pipes.

Mixing of Mass Within Storage Tanks. Most water quality models assume that the contents of storage tanks are completely mixed. The concentration throughout the tank is a blend of the tank's current contents and any water entering the tank. At the same time, the tank's internal concentration could be changing because of reactions. More refined numerical (Grayman and Clark 1993) and analytical (Boulos et al. 1995, 1998) models of non-ideal mixing in storage tanks have been proposed (Grayman and Clark 1993). These typically divide the tank into two or more completely mixed compartments and use exchange coefficient parameters to represent the movement of water between compartments.

Reactions Within Pipes and Storage Tanks. While a substance moves down a pipe or resides in storage, it undergoes reaction. The rate of reaction, measured in mass reacted per volume of water per unit of time, depends on the type of water quality constituent being modeled. Some constituents, such as fluoride, do not react and are termed *conservative*. Others, such as chlorine residual, decay with time while DBPs, such as trihalomethanes (THMs), grow with time. Some, such as chlorine again, react with materials in both the bulk liquid phase and at the liquid–pipe wall boundary.

Distribution system water quality models represent these phenomena (transport within pipes, mixing at junctions and storage tanks, and appropriate reaction kinetics) with a set of mathematical equations. These equations are then solved under an appropriate set of boundary and initial conditions to predict the variation of water quality throughout the distribution system. Even though water quality calculations are made throughout the length of each pipe, output reporting of quality is usually made only for nodes. Figure 6-1 lists the set of equations that comprises a typical water quality model.

Advective Transport in Pipes

$$\frac{\partial C_i}{\partial t} = -u_i \frac{\partial C_i}{\partial x} + r(C_i) \text{ for each pipe } i$$

Mixing at Pipe Junctions

$$C_k = \frac{\Sigma_{j\epsilon I_k} Q_j C_{j|x=L_j} + Q_{k,ext} C_{k,ext}}{\Sigma_{j\epsilon I_k} Q_j + Q_{k,ext}} \text{ for each junction } k$$

Mixing in Storage Facilities

$$\frac{\partial(V_s C_s)}{\partial t} = \Sigma_{i\epsilon I_s} Q_i C_{i|x=L_i} - \Sigma_{j\epsilon O_s} Q_j C_s + r(C_s) \text{ for each storage facility } s$$

Where

C	=	constituent concentration
Q	=	flow rate
U	=	flow velocity
R	=	reaction rate
L	=	pipe length
V	=	storage volume
I	=	set of links that flow into a node
O	=	set of links that flow out of a node

Figure 6-1 Set of equations in a typical water quality model

Steady-State vs. Extended-Period Models

Two general classes of water quality models exist—steady-state and extended-period models. Steady-state models predict the spatial distribution of water quality throughout a distribution system under the assumptions that hydraulic conditions within the system do not change, at least over the length of time needed to move water from all entry points to all exit points in the system and that storage does not affect water quality. They are simple to set up and solve, however, in most cases, their restrictive assumptions limit their applicability.

Extended-period models take explicit account of how changes in flows through pipes and storage tanks occurring over an extended period of system operation affect water quality. These models predict both spatial and temporal variations in water quality throughout a network and provide a more realistic picture of system behavior.

Computational Methods

The solution of steady-state water quality models involves solving a set of simultaneous linear equations that express the conservation of mass expressions at network nodes in terms of the unknown nodal concentrations (Wood and Ormsbee 1989, Boulos and Altman 1993). The coefficients in these equations contain information from the hydraulic behavior of the network, which is determined separately.

Several solution methods are available for extended-period water quality models (Rossman and Boulos 1996). They require that a hydraulic analysis be run first to determine how flow quantities and directions change from one time period to another throughout the pipe network. At the start of each new hydraulic time period, each pipe is divided into a number of subsegments and water is either routed between subsegments (the Eulerian approach) or the subsegments are moved downstream (the Lagrangian approach) over a succession of smaller time steps. At the same time, account is taken of any reactions occurring in each subsegment. At the end of each small time step, the water entering each node from inflowing pipes is blended together to determine a new concentration value for that node. This concentration is released from the node into its outflowing pipes at the start of the next time period. When a new hydraulic time step begins, the water quality routing begins again under a new set of hydraulic conditions.

DATA REQUIREMENTS

Hydraulic Data

As with hydraulic models, water quality models need to incorporate how a distribution system's nodes and links are connected to one another, as well as the length and diameter of each pipe. In addition, the water quality model uses the flow solution of a hydraulic model as part of its input data. A steady-state water quality model requires only a single steady-state flow value for each pipe. A dynamic water quality model utilizes a time history of flows in each pipe and flows to and from each storage tank as determined from an extended period hydraulic analysis.

Water quality models normally make no use of such hydaulic data as pipe roughness or nodal elevation, head, or pressure for water quality calculations. However, the water quality model clearly uses flows that are calculated directly in the hydraulic model.

Water Quality Data

A water quality model needs to incorporate the quality of all external inflows into the network. An example would be the concentration of chlorine leaving the clearwell of a water treatment plant as well as the quantity injected at any booster stations in the system. Extended-period models also need to include how the concentration levels of external inflows change over time. These kinds of data are obtained from existing source-monitoring records when simulating existing operations or are set to desired values to investigate operational changes.

Dynamic water quality models are supplied with a set of initial water quality values for all nodes and storage tanks throughout the network. These initial values are established in several ways. One way is to use the results of a field monitoring study if they are available. A second method is to "bootstrap" the model by running it for a sufficiently long enough period of time from an arbitrary set of initial conditions under a repeating pattern of source and demand loadings until the model reaches a repeating pattern of outputs (dynamic equilibrium). The bootstrap approach is limited by the calibration of the hydraulic model and the assumption that the demands repeat every day.

Reaction Rate Data

Chlorine Decay Coefficients. Modeling the fate of residual disinfectant is one of the most common applications of distribution system water quality models. The two most frequently used disinfectants in distribution systems are chlorine and chloramines (a combination of chlorine and ammonia). These two disinfectants have a residual that warrants modeling. The term *free chlorine* denotes the amount of uncombined chlorine that consists of both hypochlorous acid (HOCl) and hypochlorite ion (OCl-). *Combined chlorine* refers to the amount of chloramines (primarily monochloramine) present. The *total chlorine* is the sum of these two. Free chlorine is more reactive than chloramines, and its reaction kinetics were studied more extensively. Findings show that in some systems there is more loss of chlorine observed within the piping system than for the same amount of water stored in an inert container over the same period of time. This implies that there are two separate reaction mechanisms for chlorine decay, one involving reactions within the bulk fluid and another involving reactions with material on or released from the pipe wall (Vasconcelos et al. 1997).

Bulk Decay Coefficients. The rate of free chlorine decay in the bulk fluid depends on the chemical nature of the water and its temperature. Chlorine reacts rather quickly with such inorganic compounds as iron and manganese. These types of reactions are probably completed during the time spent in a treatment plant's clearwell. Reactions with organic matter proceed at a slower pace. The actual rate depends on both the amount and type of organic matter present.

Free chlorine in the bulk water phase typically decays exponentially with time, expressed in

$$C = C_o \exp(-k_b t) \tag{6-1}$$

In this equation, C_o is the initial concentration at the head of the pipe; t is the time of travel spent in the pipe; and k_b is a bulk reaction coefficient with units of 1/time.

A method for measuring k_b is described in Figure 6-2. The k_b for different waters or the same source water subjected to different treatment processes varies by more than an order of magnitude. Highly reactive waters have coefficients above 1.0/day, while slightly reactive waters show coefficients below 0.1/day (Hart et al. 1987; Vasconcelos et al. 1996). For the same water, k_b can roughly double for each 10-degree rise in water temperature. It may be necessary to assign different k_b values to different pipes throughout a distributuion system to account for the blending of source waters with different reactivities. Bulk decay coefficients in tanks deserve careful consideration, especially when daily temperature changes in a tank affect the bulk decay coefficient.

Wall Demand Coefficients. Pipe wall demand for free chlorine typically occurs in unlined metallic pipes or in pipes where a significant growth of biofilm has occurred. In the former case, the demand is caused mainly by corrosion resulting in release of iron from the pipe wall, while in the latter case, it is the organic material associated with the biofilm slime layer that exerts a chlorine demand. There are two approaches used for modeling wall demand. With the first approach, the value of k_b is increased to account for the increased chlorine demand. The second approach introduces a second rate coefficient, k_w, so that the overall coefficient K becomes (Rossman et al. 1994)

$$K = k_b + 2k_w/r \qquad (6\text{-}2)$$

The term $2/r$, where r is the pipe radius, is the pipe surface area per unit of volume available for reaction.

The wall coefficient k_w, with units of length divided by time, depends on the surface characteristics of the pipe, such as the amounts of exposed iron surface and biofilm buildup. Because older pipes tend to exhibit more chlorine demand than newer ones, and these pipes also tend to have lower hydraulic roughness coefficients, k_w is hypothesized to be inversely proportional to a pipe's Hazen-Williams C-factor. In this case the overall chlorine decay coefficient becomes

$$K = k_b + \alpha/(r\, C_{HW}) \qquad (6\text{-}3)$$

where α is a factor that is system-dependent.

Both k_w and α are very site-specific. A study of portions of four distribution systems where wall demand was being exerted found that k_w values ranged between 0.1 to 5.0 ft/day (0.03 + 1.5 m/day) and that α ranged from 10 to 650 between the different systems (Vasconcelos et al. 1996).

THM Formation Coefficients. As free chlorine reacts with organic material in water, a class of compounds known as trihalomethanes (THMs) is formed. Members of this class are suspected carcinogens and the USEPA has set a maximum limit of 80 mg/L of total THM in water and requires compliance sampling in the distribution system.

The kinetics of THM formation are rather slow, with reactions proceeding for many hours after water leaves the treatment plant. It is not uncommon for more than half the total THM production to occur in the distribution system. There is a maximum THM level that forms for each water, depending on the nature and amount of organic matter and on the chlorine level present. This amount is called the THM formation potential (THMFP). Under constant conditions of temperature and pH, THM formation with time is approximated by the following equation:

$$\text{THM} = \text{THM}_0 + (\text{THMFP} - \text{THM}_0)(1 - exp)(-k_{THM}t) \qquad (6\text{-}4)$$

where THM_0 is an initial THM concentration and k_{THM} is a formation rate constant. This constant, along with the THMFP, is determined from a laboratory test as described in Figure 6-2.

FIELD MEASUREMENTS OF WATER QUALITY PARAMETERS

Field measurements serve multiple purposes when used in conjunction with water quality modeling:

- Model development—as a means of developing water quality reaction and transformation relationships
- Model calibration—to develop the parameters used by existing water quality transformation relationships

- Model testing—to ascertain that a previously calibrated water quality model performs acceptably for the situation being studied
- Calibration and testing of the underlying hydraulic model

Three general types of field measurements are discussed in this section: water quality surveys, tracer studies, and laboratory kinetic studies. Frequently, all three studies are conducted in tandem, in order to fully parameterize a hydraulic and water quality model.

Water Quality Surveys

There are four major steps that apply to all water quality surveys: (1) development of a detailed sampling plan and preparations for sampling; (2) performance of the field sampling study; (3) post-sampling data analysis and preparation of a report documenting the field study; and (4) use of the field data in the development (parameterization) of the distribution system model.

As part of the preparation of the sampling plan, the distribution system/water quality model is applied to the sampling area using the operating conditions likely to be in effect during the field study. Based on predicted results, sampling locations and sampling frequency are established. In some cases, alternative operating conditions are simulated in order to determine their effect on the planned sampling strategy. Even though the model is not fully calibrated, and in fact, is only an approximation of the actual system, the information gained by predicting the behavior of the system under the expected sampling conditions is frequently helpful. The following issues should be addressed in developing a sampling plan, preparing for sampling, conducting the actual sampling campaign, and the postsampling analysis process (Clark and Grayman 1998).

Sampling Locations. Sampling sites should be carefully and accurately delineated in the plan. Examination of the results of the preliminary modeling runs is an important factor in selecting general locations for sampling. Samples are taken from permanent sampling taps, hydrants, public buildings, businesses, residences, or water utility facilities. The locations are selected to reflect places where water quality data is desired including water sources. Other selection criteria include accessibility throughout the entire study, required flushing times, safety issues, liability, and the degree to which the sample site is representative of some portion of the distribution system. Customer plumbing greatly affects water quality, so selecting a site close to the water main is preferable.

Sampling Frequency. Samples are taken manually or by using automated collection or analysis instruments. The frequency of sampling is generally governed by the availability of personnel, the number of samples to be analyzed in the laboratory, and the expected temporal variation in water quality in the study area. For manual sampling, a sampling *circuit* is generally established in which samples are taken sequentially by a sampling crew. The circuit should be clearly marked on a map. The time it takes to sample at a single station and over the entire circuit is estimated by a preliminary test of the circuit or by estimating the times associated with each aspect of sampling. (e.g., time to flush the sampling tap, time to take and analyze a sample, time to travel between sites, etc.). The duration of a sampling program varies depending on the goals of the study, travel time across the network, and whether or not a tracer is used in the study.

System Operation. The operation of a water system significantly impacts the resulting water quality in the system. During a sampling study, system operation conditions are viewed as a controlled set of variables rather than a random set of occurrences. This requires close cooperation between the sampling team and the systems operations group. The system operation either reflects normal operating conditions or reflects a desired set of operational conditions. Operational changes affect the time required to establish a steady state water quality condition, and therefore, the length of the sampling study.

Preparation of Sampling Sites. Prior to actual sampling, the sampling site should be adequately prepared. Preparation includes installation of faucets in hydrants, calculation of required flushing time, notification of owners, and marking the sites for easy identification. Valves and appurtenances are checked to ensure that they are in good working order.

Sample Collection Procedures. Procedures for collecting samples should be specified in the plan. These procedures include required flushing times, methods for filling and marking sample containers, listing reagents or preservatives to be added to selected samples, methods for storing samples, and data logging procedures.

Collection of Ancillary Data. Use of the sampling data in conjunction with a model requires extensive knowledge of the system. Preparations should be made to collect information during the sampling study on flows, pump operations, valve settings, and any other information required by the model.

Analysis Procedures. Samples taken in the field are analyzed at the sampling site, at a field laboratory located in the sampling area, or in a centralized laboratory. Procedures to be followed for each type of analysis should be specified in the plan.

Personnel Organization and Schedule. An important part of the sampling plan is a detailed personnel schedule for the sampling study. Ideally, the schedule includes crew assignments and a work schedule for each study participant.

Logistical Arrangements. Logistical arrangements include lodging for non-resident participants, provision for meals, transportation, and all the other minor details associated with any field study.

Safety Issues. The following safety-related concerns should be addressed in preparation for the sampling campaign: notification of police and other governmental agencies; public notification; notification of customers who are directly affected; issuance of safety equipment, such as flashlights, vests, etc.; use of marked vehicles and uniforms identifying the participants as official water utility employees or contractors; and issuance of official ID cards or letters explaining their participation in the study. Whenever possible, samplers should not work alone, especially during nighttime or in dangerous areas.

Data Recording. An organized method for recording all data should be devised including the design of a data recording sheet. Data is recorded in ink using military (24-hr) time, and the individual data sheets is numbered sequentially and transferred to a central location at frequent intervals. A date-time-location identifier on both the recording sheet and on the sample containers' labels.

Equipment and Supply Needs. Equipment includes field sampling equipment, safety equipment, laboratory equipment, etc. Expendable supplies include sampling containers, reagents, marking pens, batteries, etc. As part of the sampling plan, the needs and availability of equipment and supplies are identified and alternative sources for equipment investigated. Some redundancy in equipment is planned because equipment malfunction or loss is possible.

Training Requirements. Training of sampling crews is essential and should be specified in the sampling plan. Even though crews are composed of experienced workers, it is important that all crews be trained in a consistent set of procedures. Topics include the following: the sampling locations, sample collection and analysis procedures, data recording procedures, contingencies. All crew members should read the sampling plan prior to the training sessions.

Contingency Plans. The old adage that, "if something can go wrong during a sampling study, it will," serves as a good basis for contingency planning. Contingency planning includes equipment malfunction, illness of crew members, communication problems, severe weather, unexpected system operation, and customer complaints. Backup sampling sites are a part of a contingency plan.

Communications and Coordination. During the sampling study, key personnel are often distributed throughout the area as follows: samplers at single stations or riding a circuit, the laboratory analysts in a permanent or field lab; operations personnel at central control stations; the study supervisor at any of many locations. In order to coordinate actions during the study or to respond to unexpected events, some means of communication are needed. Alternatives include radios (normally available if you are using utility vehicles), cellular phones, walkie-talkies, or, the low-tech solution of a person circulating in a vehicle. Ideally, a study is coordinated through a central location where a study supervisor is located. The supervisor functions as the central coordinator, receiving information from the field, laboratory, and operations center and making decisions and disseminating information back to the field. Decisions on changes in sampling schedules or operating procedures are made out of this center based on full knowledge of all aspects of the ongoing study.

Calibration and Review of Analytical Instruments. A specific plan for reviewing and calibrating all instruments is necessary, and this work must be carried out prior to the study. When multiple instruments are used to measure the same parameter, each instrument is calibrated against a standard and compared. A plan to check the calibration of instruments during the study is also essential.

Preparation of Data Report. Immediately following the sampling campaign, the data is examined and transferred to a computer format using spreadsheet or database management software and a comprehensive data report is prepared. The purpose of the data report is to contain all pertinent information on the field study and subsequent lab analysis. This includes information on the sampling study, the study area, and the results of the lab analysis organized in a usable manner.

Tracer Studies

Tracer studies in distribution systems involve the observation of the movement of a substance within the distribution system. The information gained from a tracer study is used to characterize the travel times (velocity) within the system and to aid in the testing and calibration of a hydraulic model.

Tracers that are used in such studies include (1) constituents that are normally added to a distribution system, such as fluoride, which are turned off for some period and traced; (2) nontoxic, conservative substances, such as sodium chloride, calcium chloride, or lithium chloride, which are injected at a known concentration; or (3) naturally occurring substances, such as hardness, which varies between different sources.

The underlying concept behind tracer studies for use in the characterization/calibration/testing process is as follows: When modeling a conservative substance, there are essentially no water quality parameters that may be adjusted. In other words, if the hydraulic parameters are correct and the initial conditions and loading conditions for the substance are accurately known, the water quality model should provide a good estimate of the concentration of the substance throughout the network. The use of the water quality model and conservative tracer as a means of calibrating the hydraulic model is based on this relationship.

The calibration process using water quality modeling is summarized as follows:

1. An appropriate conservative tracer is identified for a distribution system. Factors that affect this decision include cost, analysis requirements, local or state regulations, and ease in handling the tracer.

2. A controlled field experiment is performed in which either (a) the conservative tracer is injected into the system for a prescribed period of time; (b) a conservative substance that is normally added, such as fluoride, is shut off for a prescribed period; or (c) a naturally occurring substance that differs between sources is traced.

3. During the field experiment, the concentration of the tracer is measured at selected locations in the distribution system along with other parameters that are required by a hydraulic model, such as tank water levels, pump operations, flows, etc. The tracer study is conducted in conjunction with a water quality survey. Depending on the selected tracer, measurement of the tracer is done manually using in-field analysis, is done in the laboratory on samples collected in the field, or involves the use of automated, continuous chemical analyzers and data loggers.

4. Standard means are used to adjust the parameters in the hydraulic model to represent the operations of the distribution system. The water quality model is used to model the conservative tracer.

5. If the model adequately represents the observed concentrations, this indicates the likelihood of a good calibration of the hydraulic model for the conditions being modeled. Significant deviation between the observed and modeled concentrations indicates that further calibration of the hydraulic model is required. Various statistical and directed search techniques are used in conjunction with the conservative tracer data to aid the user in adjusting the hydraulic model parameters so as to better match the observed concentrations.

Tracer studies were successfully conducted at many test sites around the US and in other countries (Clark et al. 1991; Elton et al. 1995; Vasconcelos et al. 1996). However, prior to embarking on tracer studies, a detailed study plan should be developed and the costs and resource requirements evaluated.

Laboratory Kinetic Studies

Laboratory tests are available for estimating the kinetics of two types of reactions represented in distribution system water quality models—bulk chlorine decay and THM formation.

Chlorine Decay Bottle Test

The chlorine decay bottle test measures the rate of chlorine decay that occurs in the bulk flow without any influence from the pipe wall. The test is conducted on the water that enters the zone of the distribution system being modeled. Waters from different sources undergo separate tests. Figure 6-2 lists the steps used to conduct the test. Results of the test provide an estimate of the bulk decay rate coefficient, k_b, that is used to model chlorine disappearance in a distribution system.

> This procedure should be followed to generate data used to estimate the rate of chlorine decay in finished drinking water.
>
> 1. Split a single sample of finished water into 10–12 amber bottles of 250 mL size or larger. Fill the bottles so they are headspace-free and cap them.
> 2. Place the bottles in a small bucket or wire cage and immerse them in a water bath in a sink with running water so that the initial water temperature of the sample can be maintained.
> 3. Starting from time zero, periodically remove a bottle and analyze the contents for free chlorine.
> 4. For each bottle analyzed, record the time (in hours from the start of the test), the free chlorine concentration of its contents, and the temperature of the water bath.
>
> An ideal schedule for analyzing the bottles would provide at least 10 chlorine values that cover a range from 100 percent down to about 25 percent of the initial chlorine concentration. A typical schedule might analyze bottles at 0, 3, 6, 12, 18, 24, 36, 48, 60, and 72 hr. However, some adjustments will have to be made if the intermediate results indicate that chlorine is decaying either much more rapidly or much more slowly than anticipated.
>
> After the test data are generated, a bulk decay coefficient is estimated by plotting the natural logarithm of chlorine concentration versus time and fitting a straight line through the points. The slope of the line is the bulk decay coefficient in units of 1/time (e.g., if time is in hours, the coefficient has units of 1/hr). When fitting the line, it is best to force it to pass through the initial measurement at time zero. Alternatively, use a nonlinear curve fitting routine, available in many commercial spreadsheet and curve-plotting software packages, to estimate kb directly from the equation
>
> $$C = Co \exp(-kbt)$$
>
> where C is the chlorine concentration measured at time t and Co is the measured concentration at time zero.

Figure 6-2 Protocol for chlorine decay bottle test

It should be possible to conduct a similar test for chloramine decay. In this case, the duration of the test is likely much longer than for chlorine because of the lower reactivity of chloramines.

THM Formation Test

A similar bottle test can be used to estimate rate parameters for the growth of THMs. This is a modified form of the standard SDS (simulated distribution system) test (*Standard Methods 1998*) where instead of a single sample being incubated for three days, multiple samples are incubated together, each for a different period of time. The test is run long enough so that THM levels plateau to a constant level. This level becomes the formation potential THMFP, and k_{THM} is estimated by plotting the natural logarithm of THM at time t minus THM at time zero against time and computing the slope of the straight line fit through these points. Alternatively, nonlinear least squares are used to estimate k_{THM} directly from the equation

$$THM = THM_0 + (THMFP - THM_0)(1 - \exp(-k_{THM}t))$$

where THM is the measured THM value at time t and THM_0 is the measured value at time zero.

MODEL CALIBRATION

Every model has a set of coefficients, parameters, or constants that reflect site-specific conditions. For distribution system models, these include pipe diameters, pipe lengths, water demands, and pipe roughness coefficients. For water quality models, they include reaction rate coefficients, source input rates, and initial water quality conditions. Some parameters are measured directly, such as pipe lengths and source input rates. Others, like pipe roughness and reaction rate coefficients, are not. Sometimes, it is possible to estimate values for such parameters by conducting field tests (e.g., hydrant pressure tests for roughness factors) or laboratory studies (e.g., bottle tests for chlorine bulk decay coefficients). The process of adjusting model coefficients to make the model's output best conform with actual observations is called calibration.

Calibration Criteria

Though many utilities and some industry organizations have developed criteria that they use in the calibration of distribution system models, at this time, there is no universally accepted criteria. Criteria in use today range from qualitative methods that rely on subjective judgments based on visual inspection of observed and modeled results to quantitative comparison of observed and predicted results utilizing statistical tests. There are advantages and disadvantages to both methods, and both have a place in the application of models.

Qualitative Criteria

A time series plot, such as is presented in Figure 6-3, is often used to compare modeled results to observed data. The modeler makes a decision as to the best match and as to whether the predicted and observed results are sufficiently close. This method relies on the experience of the modeler and the ability of the modeler to select between alternatives. Though such a method is obviously subjective, the process usually leads to acceptable criteria.

Calibration Techniques

Calibration techniques generally fall in the category of trial-and-error or ad hoc methods. Automated methods of calibration and optimization techniques have also been proposed. Calibration may be viewed conceptually as a screen showing the observed and predicted values with a large number of knobs that are adjusted in order to improve the agreement between the observed and predicted values. Each knob represents a parameter, such as the roughness coefficient, the chlorine wall demand coefficient for a pipe, or the water usage assigned to a node. Therefore, in a small distribution system, there are so many knobs, permutations, and combinations of parameter values that a systematic search becomes intractable.

Two techniques that are frequently employed are directed searches and grouped calibration. In a directed search, the modeler targets areas of significant disagreement first. For example, if a predicted high concentration at a node occurs prior to the observed similar value, parameters that affect travel time in the vicinity of the node in question, such as the roughness coefficient of upstream links, are adjusted first. In grouped calibration, a group of parameters are adjusted simultaneously. For example, if chlorine concentrations are systematically overpredicted in one area of a distribution system, the wall demand coefficient for just that area is systematically increased.

Station M003

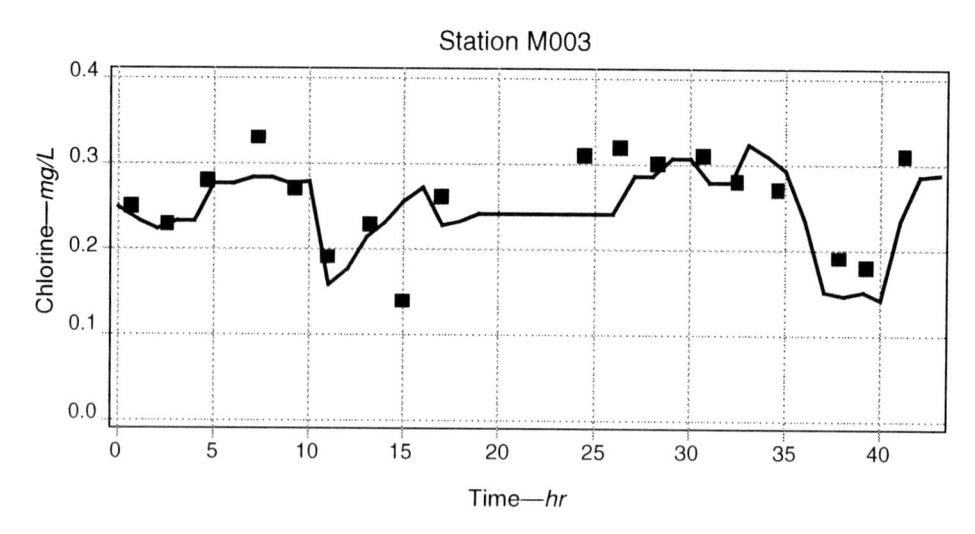

Figure 6-3 Time series plot comparing modeled results to observed data

Another example of grouped calibration is the development of a relationship between parameters followed by adjustments to all values in that group. For example, the roughness coefficient is related to pipe age and material and the assigned coefficients changed for the entire group to improve calibration. Or, the chlorine wall demand is related to the pipe roughness (inversely proportional) and the relationship varied globally to improve calibration. Because the objective of the calibration exercise is to get a good correlation between the model and the actual system, investigations into both hydraulic and water quality model parameters are warranted.

Confirmation Testing

Model parameters are adjusted so that the model faithfully reproduces observed values for the calibration data set. However, this does not necessarily guarantee that the model is applied in confidence to other situations. In order to develop this confidence, the initially calibrated model is applied to a set of data that has not been used to adjust the parameters. The degree of confidence in the model increases with the number of independent data sets that are tested.

Water quality models should always be all main models. If the smallest pipes are not included in the model, the model significantly undercalculates the actual age of water in the system.

CASE STUDY: NORTH MARIN WATER DISTRICT, CALIF.

The following case study illustrates how water quality data collected from an intensive short-term sampling program is used to calibrate both an extended-period hydraulic model and a dynamic water quality model of a distribution system. It also demonstrates some of the complexities involved in modeling chlorine decay in multisource systems where both bulk and pipe wall demands are present.

Background

At the time of the study, the North Marin Water District (NMWD) served a suburban population of approximately 53,000 people in the northern portion of Marin County, Calif. In 1993, as part of an AWWA Research Foundation project, a water quality sampling and modeling study was conducted in order to characterize and model chlorine decay in the distribution system (Vasconcelos et al. 1996).

NMWD was served by two sources of water: The Russian River, via an aqueduct used throughout the year, and Stafford Lake, which provided water from approximately May through October. Russian River water entered the system with a chlorine residual in the range of 0.3 to 0.4 mg/L and a THM concentration in the range of 10 to 20 µg/L. Stafford Lake had a high humic content and following conventional treatment and prechlorination, the water left the clearwell with a chlorine residual of approximately 0.5 mg/L and THM levels that sometimes exceeded 100 µg/L. Sodium hydroxide was added in the treatment process, resulting in average sodium levels of 23 mg/L, as compared to sodium levels in the Russian River of about 9 mg/L. The significant difference in sodium concentrations in the two sources provided an easy method for identifying the source of the water as it blended within the distribution system.

The NMWD system was divided into a series of zones and satellite systems. For the sampling and modeling study, only the primary zone (Zone I) was represented with transfers to other zones represented as external demands. Zone I contained four tanks and several pump stations used to lift water into higher pressure zones.

Sampling Study

A 42-hr sampling study was performed in July 1993 to gather data for use in the assessment and modeling aspects of the project. Prior to conducting the study, a network model was applied to determine the projected response of the system and to aid in selecting sampling stations and frequency. A skeletonized representation of the distribution system network is presented in Figure 6-4, showing the location of the sampling stations. Flows from the two sources and tank water levels were recorded every 15 min by a SCADA system.

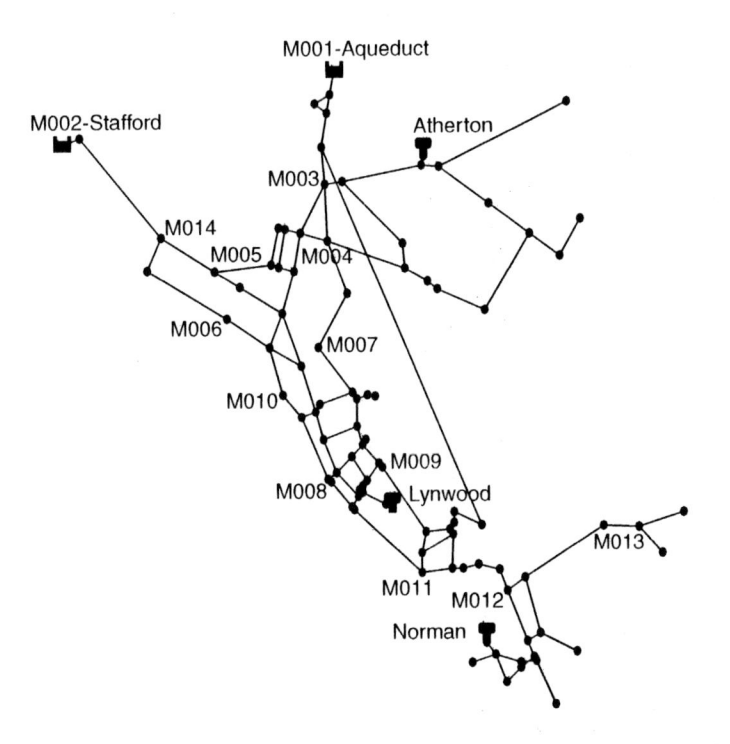

Figure 6-4 Skeletonized representation of Zone I of the North Marin Water District

During the sampling study, samples were taken at approximately 2-hr intervals at 16 stations. Each sample was analyzed for temperature, sodium, and free and total chlorine. Selected samples were also analyzed for a wide range of parameters in order to characterize the water and to study THM formation. Bottle decay tests were run to determine the chlorine bulk decay coefficient for the two sources of water and for an even mixture of the two sources. All sampling results were stored in a database management system for later graphical and statistical analysis.

Hydraulic Calibration

Following the completion of the sampling study, the initial modeling task was to test an existing hydraulically calibrated model of Zone I of the NMWD system for the conditions encountered during the study. The model was first updated to reflect recent hydraulic modifications in the network. The time history of SCADA recorded tank levels and pump station flows were used to compute total demand in the network for each hour of the sampling period. Total demands were disaggregated to individual network nodes by prorating them against demands used in the original model calibration.

Testing of the hydraulic model was done by comparing predicted and observed sodium concentrations over time throughout the system because of the blending of water from the two sources. Initial sodium levels at all nodes in the model were assigned by interpolating from the initial samples taken at the monitoring stations. During the period of sampling, the Stafford Treatment plant operated only during the day for 8- to 9-hr periods, resulting in widely fluctuating sodium concentrations throughout the system. Figure 6-5 illustrates the observed and modeled sodium concentrations at three stations: M003, which received Russian River water at all times; M005, which received Stafford Lake water when that treatment plant was operating and Russian River water at other times; and M012, which received blended water. The study relied on visual comparison (as opposed to statistical analysis) of the modeled and observed sodium concentrations, along with a comparison of the storage tank water elevation trajectories, to determine that an acceptable level of hydraulic calibration had been obtained. For systems that do not have multiple sources of water that contain significantly different water quality "signatures," similar calibration is performed by the addition of a conservative tracer, such as fluoride.

Water Quality Calibration

The sampling results were used to provide an understanding of chlorine decay kinetics in the NMWD system and to develop a calibrated chlorine model of the system. The two sources of water were very different in terms of their chlorine decay kinetics, as illustrated by their bulk chlorine reaction coefficients:

   Russian River: 1.32/day
   Stafford Lake: 17.7/day
   50/50 blend: 10.8/day

This complicated the modeling of chlorine decay within the network because most water quality models utilize a single bulk decay coefficient for each pipe that is applied to all water flowing through the pipe regardless of its source of origin.

To resolve this situation, a model run was made to determine what the average contribution of flow to each of the monitoring stations from Stafford Lake was over the 42-hr sampling period. The results, portrayed in Figure 6-6, show that the Stafford source serves mainly the western edge of the system. This suggested

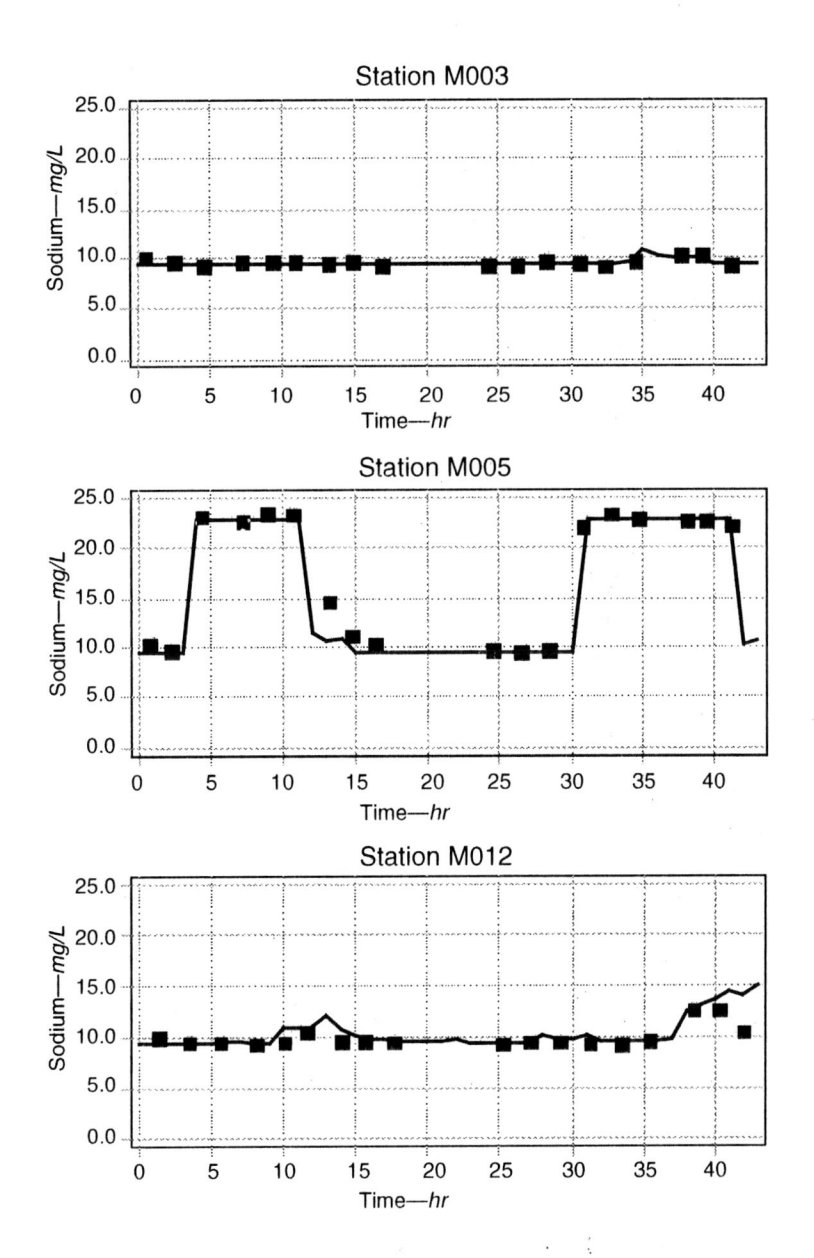

Figure 6-5 Comparison of observed and modeled sodium concentrations in the North Marin Water District

dividing the system into two zones, as shown in the figure. The pipes in the eastern zone, which receive mostly Aqueduct (Russian River) water, were assigned the bulk decay coefficient associated with the Aqueduct (Russian River) water. Because, on average, the water in the Stafford-zone pipes was close to a 50/50 blend of Aqueduct (Russian River) and Stafford water, these pipes were assigned the decay coefficient determined for the 50/50 blended water. The main line connecting Stafford to station M014 was assigned the bulk coefficient for 100 percent Stafford water because no Aqueduct (Russian River) water ever travels through this pipe. As with sodium, initial concentrations of chlorine throughout the network were assigned by interpolating from the initial sampled values at each of the monitoring stations. Inputs of chlorine from the two sources were kept at their recorded values during the duration of the sampling study.

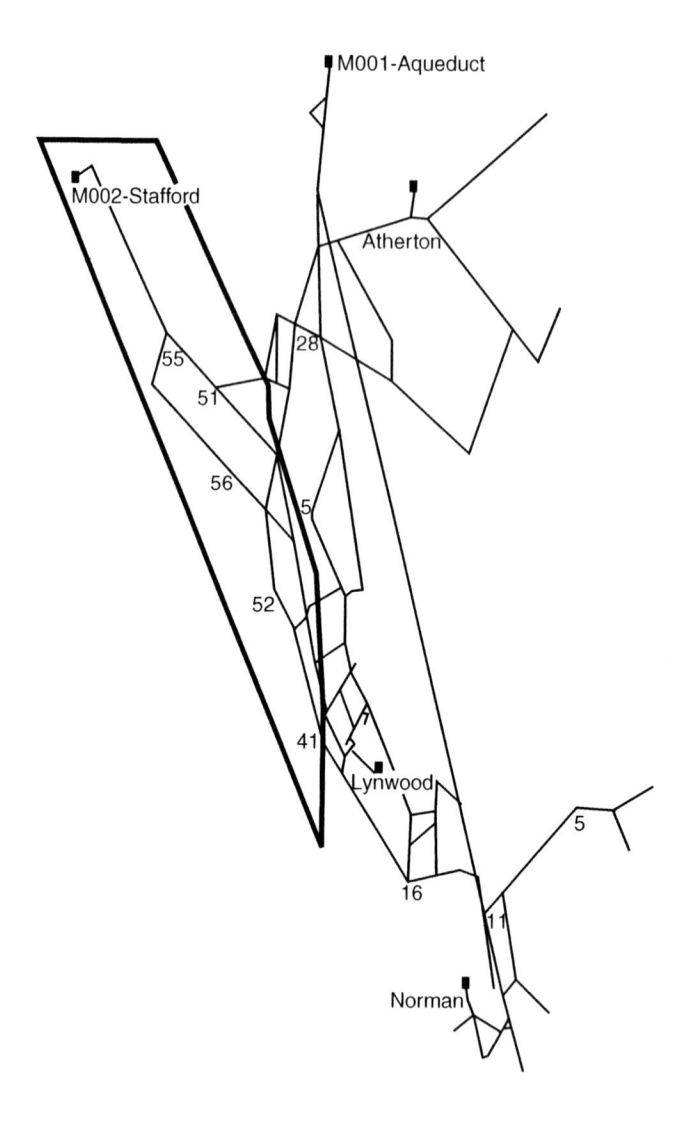

Figure 6-6 Average percent of Stafford Lake water in the North Marin Water District

Initial runs of the model were made with no pipe wall demand for chlorine. When simulated chlorine concentrations were compared with measured values, it became apparent that chlorine levels in the Stafford zone were overpredicted. This result implied that the pipes in this portion of the system were exerting an additional chlorine demand from the pipe wall. This was logically consistent with the age and material of these pipes (40–50 year-old mortar-lined steel or cast iron) when compared to the newer asbestos–cement pipes found in the southerly and easterly portions of the system.

A wall demand was then introduced into the model for the pipes in the Stafford zone of the system. The wall coefficient was systematically varied over a range of values until the overall mean absolute error between observed and predicted chlorine values at all of the sampling locations was minimized. This occurred at a wall coefficient of 5 ft/day (1.5 m/day), producing a mean absolute error of 0.05 mg/L. Time series plots comparing the model with observed chlorine levels for several of the monitoring stations are presented in Figure 6-7.

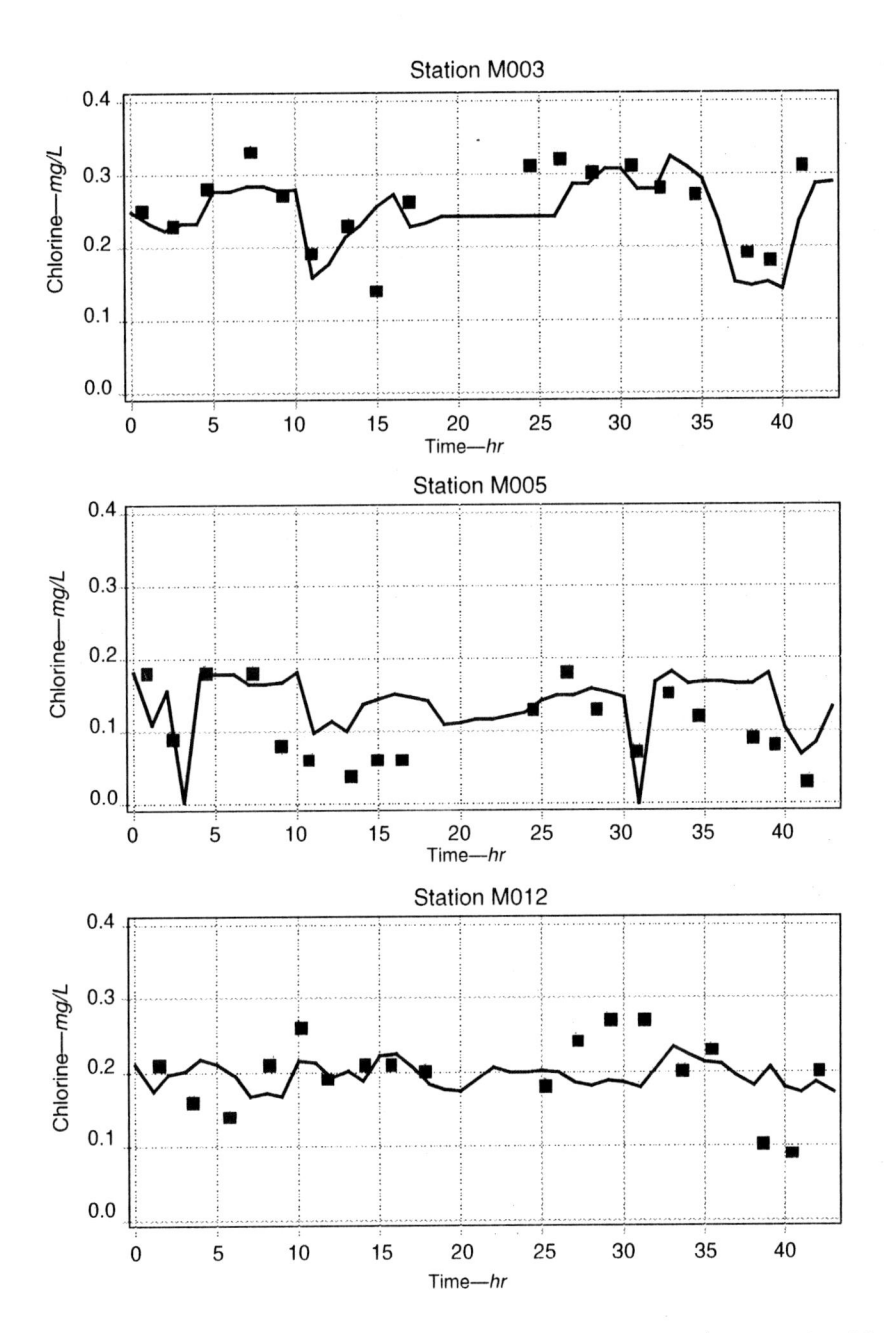

Figure 6-7 Comparison of observed and modeled chlorine residual in the North Marin Water District

This exercise demonstrated that a distribution system water quality model could be calibrated to reproduce conditions observed within the North Marin system during the July 1993 sampling period. Additional testing of the model, using data collected from other operating periods, was needed before it could be applied within the District.

REFERENCES

Clark, R.M., Grayman, W.M., Goodrich, J.A., Deininger, R.A., and A.F. Hess. 1991. Field-Testing Distribution Water Quality Models. *Jour. AWWA* 83:7:67–75

Clark, R.M., and W.M. Grayman. 1998. *Modeling Water Quality in Drinking Water Distribution Systems.* AWWA, Denver, Colo.

Grayman, W.M., and R.M. Clark. 1993. Using Computers to Determine the Effect of Storage on Water Quality. *Jour. AWWA* 85:7:67–77

Elton, A., Brammer, L.F., and N.S. Tansley. 1995. Application of Water Quality Modeling. *Jour. AWWA* 87:7:44–52

Hart, F.L., Meader, J.L, and S.M. Chiang. 1987. CLNET—A Simulation Model For Tracing Chlorine Residuals in a Potable Water Distribution Network. In *Proceedings of the Distribution System Symposium.* AWWA, Denver, Colo.

Rossman, L.A., Boulos, P.F., and T. Altman. 1993. Discrete Volume-Element Method for Network Water Quality Models. *Journal of Water Resources Planning and Management, ASCE* 119:5:505–517

Rossman, L.A., Clark, R.M., and W.M. Grayman. 1994. Modeling Chlorine Residuals in Drinking-Water Distribution Systems. *Journal of Environmental Engineering, ASCE* 120:4:803–820

Rossman, L.A. and P.F. Boulos. 1996. Numerical Methods for Modeling Water Quality in Distribution Systems: A Comparison. *Journal of Water Resources Planning and Management ASCE* 122:2:137–146

Summers, R.S., Hooper, S.M., Shukairy, H.M., Solarik, G., and D. Owen. Assessing DBP Yield: Uniform Formation Conditions. *Jour. AWWA* 88:6:80–93

Vasconcelos, J.J., et al. 1996. *Characterization and Modeling of Chlorine Decay in Distribution Systems.* American Water Works Association Research Foundation, Denver, Colo.

Vasconcelos, J.J., Rossman, L.A., Grayman, W.M., Boulos, P.F., and R.M. Clark. 1997. Kinetics of Chlorine Decay. *Jour. AWWA* 89:7:54–65

Wood, D.J., and L.E. Ormsbee. 1989. Supply Identification for Water Distribution Systems. *Jour. AWWA* 81:7:74–80

1998. *Standard Methods for the Examination of Water and Wastewater.* American Public Health Association, Washington, D.C.; AWWA, Denver, Colo.; Water Environment Federation, Alexandria, Va.

AWWA MANUAL M32

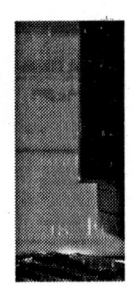

Chapter 7

Advances and Trends in Network Modeling

INTRODUCTION

This chapter discusses several topics where the mathematical approaches used in hydraulic analysis are still evolving and improving. Three topics are discussed in this section:

- Model calibration
- Tank modeling
- Optimal distribution system control (pump scheduling)

AUTOMATIC CALIBRATION

Historically, most attempts at network model calibration typically employed an empirical or trial-and-error approach. Such an approach often proves time consuming and frustrating when dealing with most typical water systems. Several researchers have proposed different algorithms for use in automatically calibrating hydraulic network models. These techniques are based on the use of analytical equations (Walski 1983), simulation models (Rahal et al. 1980; Gofman and Rodeh 1981; Ormsbee and Wood 1986; and Boulos and Ormsbee 1991), and optimization methods (Meredith 1983; Coulbeck 1984, Ormsbee 1989; Lansey and Basnet 1991; and Ormsbee, et al. 1995).

Analytical Approaches

Techniques based on analytical equations generally require significant simplification of the network through skeletonization and the use of equivalent pipes. By nature, they are also limited to applications involving a single loading or demand condition and thus do not represent the range of conditions normally encountered by the water distribution system. As a result, such techniques only provide information close to the correct results. Conversely, both simulation and optimization approaches use a complete model.

Simulation Approaches

Simulation techniques are based on solving for one or more calibration factors through the addition of one or more network equations. The additional equation or equations are used to define another observed boundary condition (such as a fire flow discharge head). By adding an extra equation, another unknown is determined explicitly. Similar to analytical approaches, the primary disadvantage of the simulation approaches is that they only handle one set of boundary conditions at a time. For example, in applying a simulation approach to a system with three different sets of observations (all of which were obtained under different boundary conditions, i.e., different tank levels, pump status, etc.), three different results are expected. Attempts to obtain a single calibration result will require one of two application strategies: a sequential approach or an average approach. In applying the sequential approach the system is subdivided into multiple zones, the number of which corresponds to the number of sets of boundary conditions. In this case, the first set of observations is used to obtain calibration factors for the first zone. These factors are fixed and another set of factors is determined for the second zone and so on. In the average approach, final calibration factors are obtained by averaging the calibration factors for each of the individual calibration applications.

Optimization Approaches

The primary alternative to the simulation approach is the optimization approach. In using an optimization approach, the calibration problem is formulated as a mathematical optimization problem consisting of a nonlinear objective function subject to both linear and nonlinear implicit bound and system constraints, as well as explicit bound constraints. Normally, the objective function is formulated so as to minimize the square of the differences between observed and predicted values of pressures and flows. The implicit bound constraints on the problem include both pressure bound constraints and flow-rate bound constraints. These conditions are used to insure that the resulting calibration does not produce unrealistic pressure or flows as a result of the model calibration process. The explicit bound constraints may be used to set limits on the explicit decision variables of the calibration problem. Normally, these variables include the roughness coefficient of each pipe and the demands at each node.

The implicit system constraints include nodal conservation of mass and conservation of energy. The nodal conservation of mass equation requires that the sum of flows into or out of any junction node minus any external demand must equal zero. The conservation of energy constraint requires that the sum of the line loss and the minor losses over any path or loop, minus any energy added to the liquid by a pump, minus the difference in grade between two points of known energy, is equal to zero. While the conservation of mass equations are linear functions of flow, the conservation of energy equations are nonlinear functions of flow, which greatly complicates the mathematical solution process.

While both the implicit and explicit bound constraints are traditionally incorporated directly into the nonlinear problem formulation, the implicit system constraints are handled using one of two different approaches. In the first approach, the implicit system constraints are incorporated directly into the set of nonlinear equations and solved using normal nonlinear programming methods. In the second approach, the equations are removed from the optimization problem and evaluated externally, using mathematical simulations (Ormsbee 1989; Lansey and Basnet 1991). This allows for a much smaller and more tractable optimization problem, because both sets of implicit equations (which constitute linear and nonlinear equality constraints to the original problem) are satisfied much more efficiently using an external simulation model (see Figure 7-1). The basic idea behind the approach is to use an implicit optimization algorithm to generate a vector of decision variables (which normally include nodal demands D and pipe roughness coefficients C that are then passed to a lower level simulation model for use in evaluating all implicit system constraints. Feedback from the simulation model includes pressures P and flows Q for use in identifying the status of each constraint, as well as numerical results for use in evaluating the associated objective function. Regardless of which approach is chosen, the resulting mathematical formulation is solved using some type of nonlinear optimization method. In general, three different approaches were proposed and used: gradient-based methods, pattern-search methods, and general heuristic methods.

Gradient-based methods require either first or second derivative information in order to produce improvements in the objective function. Traditionally, constraints are handled using either a penalty method or the Lagrange multiplier method (Edgar and Himmelblau 1988). Pattern-search methods employ a nongradient search method using objective function values only in determining a sequential path through the region of search (Ormsbee 1986). In general, when the objective function is explicitly differentiated with respect to the decision variables, the gradient methods are preferable to search methods. When the objective function is not an explicit function of the decision variables, as is normally the case with the current problem, the relative advantage is not as great, although the required gradient information is still determined numerically.

Several researchers investigated the use of heuristic methods for solving such complex nonlinear optimization problems. Perhaps the most promising of these methods is a method called genetic optimization (Goldberg 1989). Genetic optimization offers a significant advantage over more traditional optimization approaches in that it attempts to obtain an optimal solution by continuing to evaluate multiple solution vectors simultaneously. In addition, genetic optimization methods do not require gradient information. Finally, genetic optimization methods employ probabilistic transition rules as opposed to deterministic rules, which has the advantage of insuring a robust solution methodology.

Optimization Algorithm

1. Generate decision vector **X**, which satisfies the explicit bound constraints.
2. Pass **X** to simulation algorithm.
3. On return, evaluate the objective function and the implicit bound constraints.
4. Update **X** to improve the objective function while satisfying the explicit bound constraints.
5. Return to 2 and continue or stop.

$C,D \downarrow$ $\qquad$ $Q,P \uparrow$

Simulation Algorithm

1. Satisfy the implicit system constraints (i.e., the conservation of mass equations and the conservation of energy equations).

Figure 7-1 Bi-level computational framework

Genetic optimization starts with an initial population of randomly generated decision vectors. For an application to network calibration, each decision vector consists of a subset of pipe roughness coefficients, nodal demands, etc. The final population of decision vectors is then determined through an iterative solution methodology that employs three sequential steps: evaluation, selection, and reproduction. The evaluation phase involves the determination of the value of a fitness function (objective function) for each element (decision vector) in the current population. Based on these elevations, the algorithm selects a subset of solutions for use in reproduction. The reproduction phase of the algorithm involves the generation of new offspring (additional decision vectors) using the selected pool of parent solutions. Reproduction is accomplished through the process of crossover whereby the numerical values of the new decision vector is determined by selecting elements from two parent decision vectors. The viability of the generated solutions is maintained by random mutations that are occasionally introduced into the resulting vectors. The resulting algorithm is able to generate a whole family of optimal solutions and thereby increase the probability of obtaining a successful model calibration.

Trends

Distribution system model calibration is subdivided into two phases: macro-level calibration and micro-level calibration. Macro-level calibration is concerned with eliminating large errors normally associated with network geometry, measurement errors, incorrect pipe diameters, etc. Micro-level errors are normally associated with incorrect values of pipe roughness and nodal demands. With the advent and use of nonlinear optimization, it is possible to achieve some measure of success in the area of micro-level calibration. It is, of course, recognized that the level of success is highly dependent on the degree that the sources of macro-level calibration errors have first been eliminated or at least significantly reduced. While these sources of errors are not as readily identified with conventional optimization techniques, it is possible to develop prescriptive tools for these problems using expert system technology. In this case, general calibration rules are developed from an experiential database that is used by other modelers in an attempt to identify the most likely source of model error for a given set of system characteristics and operating conditions. Such a system is also linked with a graphical interface and a network model to provide an interactive environment for use in model calibration.

There is a growing advocacy for the use of both GIS technology and SCADA system databases in model calibration. GIS technology provides an efficient way to link customer billing records with network model components for use in assigning initial estimates of nodal demands (Basford and Sevier 1995). Such technology also provides a graphical environment for examining the network database for errors. One of the more interesting possibilities with regard to network model calibration is the development and implementation of an on-line system model by linking the model with an on-line SCADA system. Such a configuration provides the possibility for a continuing calibration effort in which the model is continually updated as additional data is collected through the SCADA system (Schulte and Malm 1993). Finally, Bush and Uber (1998) developed three sensitivity-based metrics to rank potential sampling locations for use in model calibration. Although the documented sampling application was small, the approach the authors developed provides a potential basis for selecting improved sampling sites for improved model calibration.

Practical Aspects of Optimal Calibration

All calibration is susceptible to problems caused by errors in measurement. However, optimal calibration lacks the ability to disregard or discount the importance of questionable data and leads to "optimal" but incorrect calibration. Optimal calibration methods are quite good at matching field measurement of head or flow. However, there are thousands of unknowns in model calibration and usually only a handful of observations, such that there are an infinite number of possible good solutions (in terms of roughness and demand) to any calibration problem, which only differ by the error in field measurements (see chapter 3).

Calibration problems are classified as *underspecified,* in that there are far more unknowns than equations (independent observations). Some of this problem is reduced by lumping parameters into groups. For example, all 70-year-old cast iron pipe has the same C-factor. Grouping reduces the time it takes to find a solution and minimizes unjustifiable values for parameters simply for the sake of optimization but is based on the assumption that the grouping is correct (e.g., pipe in a group does have the same roughness). The tradeoffs in this process are discussed by Mallick et al. (2002).

The problems caused by the underspecified calibration problem are minimized by collecting additional field data. These additional data points must be independent observations. Measuring the same pressure every hour for a day does not produce 24 measurements unless conditions change dramatically between measurements (e.g., different pumps operating).

Before using any data for optimal model calibration, that data should be screened to ensure that it provides useful information for the problem. As stated in chapter 3, a test for any data used in calibration is

$$\text{Head loss} \gg \text{error in measurement in head loss}$$

This rule is based on the principle that head loss measurements are not used if the head loss is indistinguishable from errors in measurement.

Because optimal calibration solutions determine parameter values (e.g., roughness, demand) that force field measurements to match model results, a single incorrect value entered for a field observation provides the optimal solution as a solution that matches the incorrect data but is misleading in terms of roughness and demands.

Optimal calibration does not eliminate the need for modelers to understand hydraulic principles and the operation of their system.

TANK MODELING

Tanks and reservoirs are traditionally designed, and subsequently operated, to meet hydraulic objectives: provide emergency storage, equalize pressure in a system, and balance water usage throughout a day (Boulos et al. 1996). However, studies have indicated that water quality and mixing issues must also be considered in the design, operation, and maintenance of storage facilities. For example, evaluations of outbreaks of waterborne diseases point to storage facilities as having a likely role in the incidents (Geldreich 1996), and several water quality studies have documented the potential negative impacts of tanks and reservoirs on the water quality in and leaving the facilities (Clark and Grayman 1998).

Models of tanks and reservoirs provide a means of addressing many questions concerning the effects of design, operation, and maintenance of storage facilities on water quality. Two important questions are

"How do we determine whether a proposed or existing tank or reservoir is negatively impacting water quality?"

"How should we design new storage facilities, modify existing facilities, and operate these facilities in order to minimize negative water quality impacts?"

Types of Models

The application of models to represent finished water tanks and reservoirs is a relatively new development and one that is still actively evolving (Grayman et al. 1999). The types of models that are used to represent tanks and reservoirs are shown in Figure 7-2. Historically, both mathematical and physical scale models were applied. Generally, the advent and development of computers popularized the use of mathematical models but also strengthened the capabilities in scale modeling through improved model controls and data collection procedures. Both of the general classes of models have advantages and disadvantages that should be evaluated prior to selection of a particular modeling strategy.

The options within the field of mathematical modeling of water tanks and reservoirs are quite broad and fall within two general areas: hydrodynamic models and systems models.

Computational Fluid Dynamics

Hydrodynamic models utilize equations describing the actual movement of water within the storage facility and the physical, chemical, and biological processes affecting water quality constituents in the facility. Such models are based on the principles embodied within the Navier-Stokes equations. Hydrodynamic models of the type discussed in this section are known as computational fluid dynamics (CFD) models. In a CFD model, the facility being modeled is represented by a mesh grid. Equations are written for each cell to describe the fluid movement into and out of them. Factors included in the equations are geometry, convection, and quasi-random processes (diffusion, heat-conduction, viscous action, turbulence, etc.). Thus, results must be carefully interpreted. CFD models are an effective tool in simulating both distribution system reservoirs (Hanson and Boulos 1997, Grayman and Kirmeyer 1999) and clearwells (Hanson et al. 1998, Crozes et al. 1999).

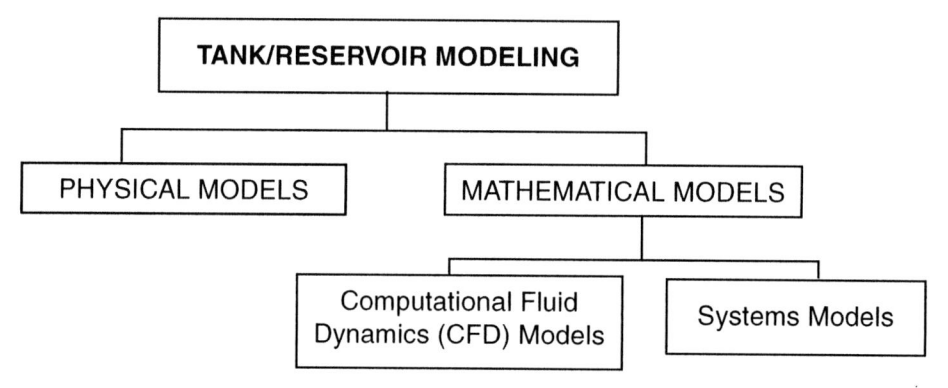

Figure 7-2 Types of tank/reservoir models

Systems Models

Systems models differ significantly from hydrodynamic models in that they treat the physical system essentially as a "black box" with primary emphasis placed on developing a relationship between the inputs to and subsequent outputs from the tank. In this type of model, the actual processes occurring within the tank are secondary in importance and are usually represented by statistical or empirical relationships that represent the interior processes in a highly conceptual manner. The simplest and most commonly used systems model of a tank is the completely mixed model, which assumes that the tank is instantaneously and completely mixed at all times. Currently, commercially available distribution system models represent tanks as completely mixed systems.

Other elemental systems models of tanks include the following: plug flow models, which assume a "first in–first out" behavior and are most representative of baffled clearwells; short-circuiting models, which assume a "last in–first out" behavior; and stagnant zone models, which recognize that a tank is not completely mixed, and certain portions of the tank experience less interchange of flow with the main body of the tank. The systems models are also used to represent nonconservative constituents by adding a first-order decay function and are used to track the water age leaving the tank. For example, the plot in Figure 7-3 shows the long-term variation in age of water leaving a reservoir and illustrates the seasonal variability in this situation.

Basic elemental systems models are combined to develop a comprehensive systems model that represents the tank as a series of conceptual compartments, such as the 3-compartment model (Boulos, Nas and Bowcock 1996, Grayman et al. 1996), with the interaction of compartments representing the short-circuiting and stagnant zone concepts. Compartments are arranged in other patterns to emphasize or represent other phenomena. The compartment models are generally parameterized by trial-and-error fitting to field data of time series of inflow and outflow concentrations. CompTank (Grayman et al. 1999) is a user-friendly systems model that provides a wide range of alternative elemental and compartment methods.

Physical Scale Models

Physical scale models were important tools in water resources studies for many years and before computers served as the primary means of developing simulations of complex hydraulic phenomena. The principles of scale modeling are well established based on the concept of similitude. In addition to issues of similitude, there are pragmatic issues of controlling the hydraulic input conditions and recording the results of the modeling. With smaller scale models, fine variations in flows, time, and dimensions correspond to significantly larger variations in the real world system. Thus, significant care must be exercised to assure accurate measurements in the scale model.

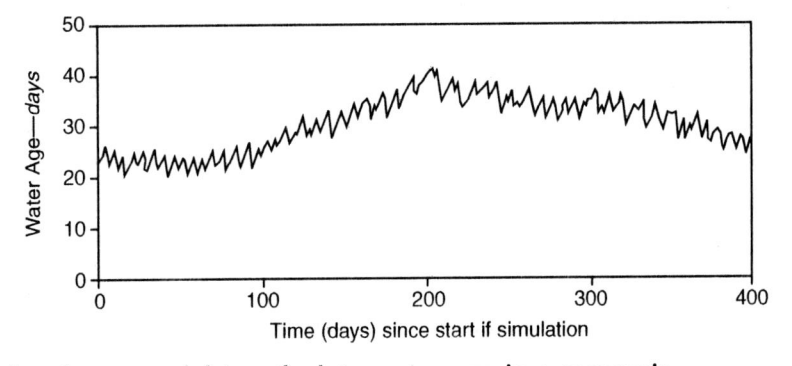

Figure 7-3 Use of systems model to calculate water age in a reservoir

The results of simulations are observed visually, recorded through optical means, or measured through digital or analog sensors. Visual observations, such as viewing the movement of dye within a model while timing its movement with a stopwatch, provides a general qualitative understanding of the model results but provides only limited quantitative information. Photographing or videotaping the results through a transparent side or open top and later analysis of the pictures through manual or computerized means provides an effective nonintrusive mechanism for quantitative analysis of the model results. However, it is difficult to capture results in thee dimensions using this method. Use of a tracer, such as lithium chloride or sodium chloride injected into the inflow stream and tracked through use of electrodes or grab samples provides a better means for quantitative measurements (Rossman and Grayman 1999). When sensors are placed within the model itself in order to capture the data (i.e., flow, temperature, concentrations, conductivity) in a spatially arrayed manner, care must be exercised to assure that the sensors do not significantly affect the hydrodynamic phenomena being modeled.

Comparison of Modeling Alternatives

The three general types of models, systems models, hydrodynamic models, and physical models, are useful techniques for simulating the behavior of tanks and reservoirs. However, each of these models has certain advantages and highly differing costs. Therefore, the selection of one or more of these types of models for a particular application involves an examination of the specific needs and the benefits and costs of the various models types.

Systems models are attractive for their simplicity, low cost, flexibility, adaptability to use with distribution system models, and speed of computation. A wide range of alternative parameter sets are tested in a few hours to determine the best-fit values. The most limiting disadvantage is the difficulties associated with parameterizing the model. Because the model parameters are not directly measured in the field, they must be derived by performing best-fit analysis on a set of field-measured inflow and outflow concentrations. This means that the model is not effectively applied to a tank or reservoir prior to its construction unless the model is applied and verified for similar situations.

Hydrodynamic models are, by nature, significantly more complex than systems models. The primary advantage of hydrodynamic models over systems models is in the parameterization. Because hydrodynamic models represent the underlying physical processes, they are applied with limited field data. As a result, hydrodynamic models are used to study mixing and water quality in tanks or reservoirs prior to their construction or modification and under a wide range of operating and seasonal conditions. The disadvantages of the greater complexity are higher costs to acquire or construct the models, need for more sophisticated hardware, increased training in application of the models, and significantly increased computer simulation time. As a result of the greater required simulation times, it is infeasible to directly incorporate a hydrodynamic-based tank model into a network simulation program. However, a hydrodynamic model could be used as a mechanism for parameterizing a systems model, which in turn could be part of a distribution system model.

Physical scale models are historically the domain of universities, research institutions, and specialized companies. They are usually built to represent a specific tank or reservoir and are not available off the shelf. Like the hydrodynamic model, they are built without having limited field data so they are appropriate for modeling proposed facilities, testing alternative operating policies, and studying the effects of retrofitting programs. Costs vary significantly from a few thousand dollars for small-scale qualitative models to much more for detailed, larger scale models. The disadvantage of scale models is the significant initial setup time associated with the model with a network simulation model. The adjustable models were constructed and readily available. For example, a cylindrical basin with movable inlets and outlets is used to simulate a very wide variety of ground level and elevated tanks. Scale models are also utilized to parameterize systems models or as a means of testing hydrodynamics models.

Conclusions

The field of modeling mixing and water quality characteristics of tanks and reservoirs is still developing. There are advantages and disadvantages to each of the model types, and an understanding of these is essential prior to the selection or development of a particular type of modeling. It is expected that as models of tanks and reservoirs are used more routinely, the availability and ease of use of the models should increase accordingly.

OPTIMAL CONTROL OF PUMPS IN WATER DISTRIBUTION SYSTEMS

Water distribution systems are controlled to provide hydraulic performance, water quality performance, and economic efficiency. Measures of hydraulic performance include pressure levels, fire protection, water quality, and various measures of system reliability. Measures of water quality performance include water age and disinfection residual levels. Economic efficiency is influenced by such factors as general operation and maintenance costs and pumping costs. In conventional water supply systems, pumping of treated water represents the major portion of the total energy budget. In groundwater systems, the pumping costs normally represent the major portion of the total operating cost. Therefore, most optimal control strategies for water distribution systems have focused on minimizing such operational costs. The following discussion applies to networks where there is a storage reservoir downstream of the pump stations to be optimized. This storage capacity is utilized to pump at times when energy costs are lower. Without a storage reservoir, the pumps must operate to meet water demand. In most water distribution applications, the required volume of water must be lifted the necessary number of feet, so the savings from optimal pump scheduling is marginal.

With respect to the minimization of operational costs, the purpose of an optimal control system is to provide the operator with the least-cost operation policy for control units (e.g., pump stations, booster chlorinators, etc.) in the water-supply system. The operation policy for a system is simply a set of rules or a schedule that indicates when a particular control unit or group of control units is turned on or off over a specified time period. The optimal policy results in the lowest total operating cost for a given set of boundary conditions and system constraints.

When considering the operation of a water distribution system, at least three different levels of automated control are identified. The lowest level of control is the computer-monitored command structure, or real-time monitoring. At the heart of this arrangement is a computer that collects and logs data, monitors the operational status of the system, and transmits an operator's directives to various control devices in the field. The next level of automated control is the computer-assisted command structure, sometimes called real-time simulation. This control structure provides operators with an interactive environment incorporating a SCADA system linked with software capable of predicting the state of the hydraulic system. The computer-assisted control configuration enables operators to examine the consequences of their actions before the actions are actually implemented in the field. The final level of control is the computer-directed command structure. The computer-directed structure is often called real-time control because the structure is able to provide an operating policy, possibly an optimal one, based on parameter forecasts and to implement the policy automatically via a direct link between the central processing computer and field-sited control devices. As a safeguard, the operator has complete manual override of the control structure at all times (Caves and Earl 1979).

Both the computer-assisted and computer-directed systems contain three major components in addition to the associated SCADA system: a hydraulic network model, a demand forecast model, and an optimal control model. Each of these components is discussed in the following sections.

Hydraulic Network Models

To evaluate the cost of a particular pump-operating policy or to assess the associated operational constraints, some type of mathematical model of the distribution system is required. Potential model structures include mass balance, regression, simplified hydraulic, full hydraulic simulation, and the use of artificial neural networks.

Mass Balance Models. Mass balance models are normally restricted to systems that contain a single storage tank. In a simple mass balance model of a single-tank system, the flow into the system equals the demand plus the rate of change in storage in the tank. The pressure head requirements to achieve the flow into the tank are neglected, and it is assumed that a pump combination is available that achieves the desired change in storage. Nodal pressure requirements are commonly assumed to be satisfied if the tank remains within a desired range. When using a mass balance model, care must be taken when determining the cost to pump a given flow because the operating cost is related to both the discharge and energy added to the flow.

In addition to use of a mass balance approach for single-tank systems, multidimensional mass balance models are also developed. These models consist of weighted functional relationships between tank flow and pump station discharge. The weights associated with the functional relationships are determined using linear regression (Sterling and Coulbeck 1975a) or linearization of the nonlinear network (Faliside and Perry 1975).

The main advantage of mass balance models is that the system's response is determined much faster than from simulation models. Thus, these models are well suited for use with optimization strategies that require large numbers of simulation analyses (Joalland and Cohen 1980). In general, mass balance models are more appropriate for regional supply systems in which flow is carried primarily by major pipelines rather than by distribution networks in which the hydraulics are commonly dominated by looped piping systems.

Regression Models. Instead of using a simple mass balance model, the nonlinear nature of the system hydraulics is represented more accurately by using a set of nonlinear regression equations. Information required to construct such models is obtained in a variety of ways. Regression curves are generated by repeated execution of a calibrated simulation model for different tank levels and loading conditions (Ormsbee et al. 1987) or by the use of information from actual operating conditions to form a database relating pump head, pump discharge, tank levels, and system demands (Tarquin and Dowdy 1989).

Regression models have the advantage of being able to incorporate some degree of system nonlinearity while providing a time-efficient mechanism for evaluating system response. However, regression curves and databases contain information only for a given network over a given range of demands. If the network changes appreciably or if forecasted demands are outside the range of the database, such an approach provides erroneous results. Moreover, regression curves are approximations of the system's response. Unless the curves are close approximations of the actual response, errors accumulate over the course of operation that adversely affect the optimization algorithm and the accuracy and acceptability of its results.

Simplified Network Models. As an intermediate step between a nonlinear regression model and a complete nonlinear network model, simplified hydraulic models are used. In such cases, the network hydraulics are approximated using a macroscopic network model or are analyzed using a system of linearized hydraulic equations. Macroscopic models represent the system by using a highly skeletonized network model. Typically, only a pump, a lumped resistance term (a pipe), and a lumped demand are included. DeMoyer and Horowitz (1973) and Coulbeck (1984) used macroscopic models with multiple terms relating the effect of various system components, but in a single equation.

In certain cases (i.e., where the system boundary conditions are essentially independent of pump-station discharge), it is possible to represent the system hydraulics using a simple linear model. Jowitt and Germanopoulos (1992) appropriately used an approximate linear model for a system dominated by large pump heads. In this case, small variations in tank levels did not have a significant impact on pump operations. In a similar application, Little and McCrodden (1989) developed a simple linear model for a supply system where the bead in the controlling tank was held constant. The coefficients for both model types are determined after extensive system analysis. As a result, such models must be evaluated on a system-dependent basis to judge their acceptability.

Full Hydraulic Simulation Models. Network simulation models provide the capability to model the nonlinear dynamics of a water distribution system by solving the governing set of quasi-steady-state hydraulic equations. For a water distribution system, the governing equations include conservation of mass and conservation of energy. These equations are solved in terms of adjustment factors for junction grades (Shamir and Howard 1968), loop flow rates (Epp and Fowler 1970), and pipe flow rates (Wood and Charles 1972).

In contrast to mass balance and regression models, simulation models are adaptive to both system changes and variations in spatial demands. For example, if a tank or large main is suddenly taken out of service, a well-calibrated simulation model still provides the hydraulic response of the modified system. A mass balance or regression model, on the other hand, requires modification of the database or regression curves to account for the changes in the system's response. Although simulation models are more robust than either mass balance or regression models to simulate a distribution system, they generally require more data to formulate. They also require a significant amount of work to calibrate properly. Because such models require a greater computational effort than either mass balance or regression models do, they generally are more useful with optimal control formulations that require a minimum number of individual system evaluations.

Neural Network Models. To reduce the computational requirements of a full hydraulic simulation model, the model is replaced with an artificial neural network representation of the system's response (Ormsbee and Lingireddy 1995b). In this case, the neural network is completely trained off-line and used instead of the network model. The data required to train the neural network comes from multiple applications of a previously calibrated hydraulic simulation model. Alternatively, the neural network is trained on-line using real time or archived data obtained from a SCADA system.

Neural networks comprise a set of highly interconnected but simple processing units, each responsible for carrying out only a few rudimentary computations. When provided with a sequential set of inputs and outputs for a given system, the network organizes itself internally in a way that allows it to reproduce an expected output for another given input. The internal process of self-organization or developing generalized representation of the system is referred to as the training process and is crucial for the efficient reproduction phase of the neural network. A neural network is said to be well-trained if the deviation between the output from the neural network and the specified output is within a tolerable limit. On the basis of the network topology, node characteristics, and learning process, several types of neural networks can be developed.

Demand Forecast Models. To develop an optimal pump operating policy, network system demands must be known. Because the actual daily demand schedule for a municipality is not known in advance, the optimal operating policy is estimated using forecasted demands from a demand-forecast model. Forecasted demands are incorporated into the optimal control model using a lumped, proportional, or distributed approach. In a lumped approach, system demands typically are represented by a single lumped value. Such an approach is normally used in conjunction with mass balance hydraulic models. Proportional demand models are normally used in conjunction with regression-based hydraulic models. In such instances, regression relationships are derived from a single demand pattern that may vary proportionally to the total system demand. A distributed demand approach is applicable when using a full network simulation model. In such an approach, the total system demand is distributed both temporally and spatially among the various network demand points. Such an approach enables the development of optimal control policies that are adaptable to significant variations in system demand that occur over the course of the designated operating period.

Distributed demand forecast models typically employ three steps: (1) they predict the daily demand; (2) they distribute the daily demand spatially among the junction nodes; and (3) they distribute the junction demands temporarily over a 24-hr operating time horizon. Prediction of the daily demand is accomplished by considering such factors as daily weather conditions, weather forecasts, seasons of the year, and past trends in water use (Maidment et al. 1985; Moss 1979; Ormsbee and Jain 1994; Sastri and Valdes 1989; Smith 1988; Steiner 1989). Distribution of the daily system demand among the junction nodes is accomplished using past meter records or real-time database information. Disaggregation of daily junction demands into smarter time intervals is accomplished by considering the day of the week and seasonal patterns of diurnal demand (Bree et al. 1976; Chen 1988a; Coulbeck et al. 1985; Perry 1981).

Techniques for estimating demand are generally available but the availability of data (both spatial and temporal) has limited the development and application of many available tools. As a result, additional work is still needed in this area, including better methods for short-interval prediction and spatial desegregation using historical short-term data. With an increase in the availability of comprehensive SCADA databases, improved model formulations, such as artificial neural networks, performance is expected to be attainable.

Control Models. Proper selection of the optimization algorithm for use in solving the associated control model often means the difference between a sluggish or even nonperforming control model and one that functions extremely well. The choice of an appropriate optimization algorithm is governed by the characteristics of the problem to be solved. Several mathematical programming techniques, such as linear programming (LP), dynamic programming (DP), and nonlinear programming (NLP), are available to solve the optimal control problem (Reklaitis et al. 1983). By far, DP was the optimization algorithm of choice by past researchers. Typically, DP was used in an implicit control formulation with tank water level generally serving as the control variable. When DP is used, the control problem is broken down into a series of discrete time steps (stages) with a prescribed set of potential control variable values (states). The optimal solution to the control problem is found by evaluating all state transitions between adjacent stages as opposed to evaluating all state transitions between all stages (i.e., total enumeration). By evaluating the state transitions between individual stages, a complex problem involving multiple subproblems is reduced to a series of problems involving a single variable. The main problem associated with the use of DP is the "curse of dimensionality," in which the computational efficiency of the method significantly decreases as the number of control variables increases. Attempts to circumvent this problem relied on the use of spatial decomposition schemes or the recasting of the problem in terms of alternate decision variables and solving using other mathematical programming techniques.

LP is the branch of mathematical programming that is used to solve problems where the objective function and all constraints are linear functions of nonnegative decision variables (Murtagh 1981). Nonlinear problems are frequently solved via LP by assuming that portions of the object function and constrained solution space are approximately linear within a prescribed interval. LP problems are solved using an approach called the simplex method, originally developed by Dantzig in the late 1940s. The simplex method offers an efficient means of finding the optimum solution of a linear optimization problem by repeatedly selecting the decision variable that causes the greatest improvement in the objective function. As a result of the nature of the linear solution space, the optimal solution of an LP problem always lies at the intersection of two or more constraints. The simplex method uses this feature of convex problems to its advantage by traveling along constraints to the intersection of other constraints. Once an initial feasible solution is determined, the algorithm identifies an adjacent point that improves the objective function and moves along a constraint to the new point. By examining the gradients of each constraint passing through the current point, a new point is selected and the process is repeated until the optimal solution is found.

The third type of control model uses NLP. As the name implies, NLP is useful for problems where the objective function or the constraints of an optimization problem, or both, are nonlinear. Unlike LP and DP, NLP involves a large number of different techniques used to solve an optimization problem. Such techniques range from elaborate gradient-based techniques (Lasdon and Waren 1986) to conceptually simple direct search methods (Reklaitis et al. 1983). Several researchers are investigating the use of more heuristically based methods, such as simulated annealing (Kirkpatrick 1983, 1984; Goldman, F.E. 1998) and genetic algorithms (Ormsbee and Lingireddy 1995b).

The Optimal Control Problem

The optimal control problem for a water distribution system is expressed in terms of a set of decision variables (the things to be varied or controlled), an objective function (an equation written in terms of the decision variables that quantifies the objective, e.g., cost), and constraints that represent restrictions on the values that the decision variables assume. Mathematically the problem is expressed as

$$\text{Minimize } F(X_1, X_2, \ldots X_n) \quad (1)$$

$$\text{Subject to: } G(X_1, X_2, \ldots X_n) = 0 \quad (2)$$

$$H(X_1, X_2, \ldots X_n) > 0 \quad (3)$$

$$X_H > X_1, X_2, \ldots X_n > X_L \quad (4)$$

Where $F(X_1, X_2, \ldots X_n)$ is the objective function written in terms of a set of n decision variables; $G(X_1, X_2, \ldots X_n) = 0$ represent several implicit system constraints; $H(X_1, X_2, \ldots X_n) > 0$ represents several implicit bound constraints; and $X_H > X_1, X_2, \ldots X_n > X_L$ represents several explicit bound constraints. In the following discussion, the optimal control problem is illustrated by considering the problem of optimal pump control.

Decision Variables. The optimal control problem for a water supply pumping system is formulated using either a direct or an indirect approach, depending on the choice of the decision variable. Direct formulation of the optimal control problem divides the operating period into a series of time intervals. For each time interval, a decision variable is assigned for each pump, indicating the portion of time the pump is operating during the time interval. The objective function for the control algorithm is composed of the sum of the energy costs associated with the operation of each pump for each time interval. The problem is then solved using either LP or NLP (Chase and Ormsbee 1989; Jowitt et al. 1988; Ormsbee and Lingireddy 1995a, 1995b). The pump-control policy that results is classified as explicit (or discrete) because the policy is composed of the required pump combinations and their associated operating times.

Instead of formulating the control problem directly in terms of pump operating times, the problem is expressed indirectly as a surrogate control variable. Such cost relationships are developed from multiple regression analyses of actual cost data or from the results of multiple mathematical simulations of the particular system. When tank level is used as the surrogate control variable, the objective becomes one of determining the least cost tank-level trajectory over the specified operating period. When pump station discharge (or pump head) is used as the control variable, the objective is to determine the least cost time distribution of flows (or heads) from all the pump stations. The pump control policies that result from such formulations are classified as implicit (or continuous) because the individual pump operating times associated with the optimal state variables are not determined explicitly (Fallside and Perry 1975; Sterling and Coulbeck 1975a; Zessler and Shamir 1989). However, the set of state variables associated with such an implicit solution normally is converted into an explicit (discrete) policy of pump operating times by subsequent application of a secondary optimization program (Coulbeck et al. 1988b; DeMoyer and Horowitz 1975; Lansey and Awumah 1994).

The Objective Function. The operating cost for a pumping system typically is composed of an energy consumption charge and a demand charge. The energy consumption charge is the portion of the electric utility bill based on the kilowatt-hours of electric energy consumed during the billing period. The demand charge represents the cost of providing surplus energy and usually is based on the peak consumption of energy that occurs during a specific time interval. The majority of existing control algorithms for water distribution systems only consider energy-consumption charges. This is primarily the result of the wide variability of demand-charge-rate schedules and that the billing period for such charges vary between one week and one year. When such charges are not explicitly included in the optimal control objective function, they are either ignored or are addressed via the system constraints.

When the demand charges are excluded from the objective function, the objective function is expressed solely in terms of the energy-consumption charge. In general, energy-consumption charges are reduced by decreasing the quantity of water pumped, decreasing the total system head, increasing the overall efficiency of the pump station by proper selection of pumps, or using tanks to maintain uniform, highly efficient pump operations. In most instances, efficiency is improved by using an optimal control algorithm to select the most efficient combination of pumps to meet a given demand. Additional savings are achieved by shifting pump operations to off-peak water demand periods through proper filling and draining of tanks. Off-peak pumping is particularly beneficial for systems operating under a variable electric rate schedule.

Operational Constraints. Constraints associated with the optimal control problem consist of physical system limitations, governing physical laws, and externally defined requirements. Physical system constraints include bounds on the volume of water stored in tanks, the amount of water supplied from a source, and valve or pump settings. The physical laws related to a supply and distribution system are the conservation of flow at nodes (conservation of mass) and conservation of energy around a loop or between two points of known total grade. Also included in this set are relationships between head loss and discharge through a pipe, pump, or valve. Typically, the only external requirements are to meet the defined demands and to maintain acceptable system pressure heads. Pressure-head requirements have both upper and lower bounds to avoid leakage and ensure satisfying user requirements. Additional constraints are added to restrict the tank levels to stay within a preset range of values.

When solving the optimization problem, the system's state at the time of analysis is known, and an assumed final condition is set as a target. The initial state of the system includes the pump operations and tank levels, whereas the final state defines the end-of-cycle tank levels. The period of analysis usually is a one-day cycle, although longer periods can be considered. The cycle for most control schemes typically begins with all tanks either completely full, or at a preset lower level, and ends 24 hr later with the same condition (Shamir 1985).

Although not normally considered explicitly in most control algorithms, it should be recognized that pump maintenance costs constitute a significant secondary component of any pump operation budget. Pump wear is directly related to the number of times a pump is turned on and off over a given life cycle. As a result, operators attempt to minimize the number of pump switches while simultaneously determining least cost operations. This problem is not as significant for newer pumps, which are better designed and made of more durable materials, but it is a major concern in many older systems. Unfortunately, sufficient data are not currently available to permit the incorporation of such costs directly into the objective function. Instead, limits on pump switches normally are set through the use of the system constraints (Lansey and Awumah 1994) or an approximate cost term (Coulbeck and Sterling 1978).

SUMMARY

Many researchers have developed optimal control formulations to minimize the operating costs associated with water supply pumping systems. For a more in-depth review the reader is referred to chapter 16 in the *Water Distribution System Handbook* (AWWA-McGraw Hill 2000) or an earlier review by Ormsbee and Lansey

(1994). The choice of the appropriate algorithm for a particular application depends largely on the physical characteristics of the system. The most straightforward approach for single-tank systems is a formulation with tank level as the state variable in a DP model. Such an approach is generally efficient when the system demands are lumped at a single node or are assumed to vary proportionally. Attempts to incorporate the impact of the spatial variability of demand or changes in the operational status of various system components normally requires the use of an alternative formulation. For systems that contain a reasonable number of pumps, it is plausible to use a pump run-time model (Chase and Ormsbee 1991; Ormsbee and Lingireddy 1995a, 1995b). When the total number of pumps is large, the use of an implicit pump station decision variable may be more appropriate (Lansey and Zhong 1990).

For multisource/multitank systems that are highly serial or permit a convenient subdivision into distinct hydraulic units, a dynamic programming spatial decomposition approach is feasible. However, for systems that do not readily permit spatial decomposition, control algorithms normally require lumped pump-station models and the pump run-time models to accommodate directly the nonlinear dynamics of most multisource/multitank systems that makes the use of nonlinear optimization an acceptable trade-off. As more tanks and distributed demands are considered, a more detailed simulation model is necessary. The trade-off is between optimization time requirements, accuracy, and the precision of the associated hydraulic model. Typically, these trade-offs are evaluated on a network-by-network basis because rules of thumb are difficult to derive.

When using pump station discharge as a surrogate control variable, the selection of a discharge-cost relationship is made with extreme care. In most cases, pump station discharge varies with both demand and tank level. As a result, the associated cost and hydraulic relationships must have two independent variables (demand and tank level) (Ormsbee et al. 1989), or they must account for the required pressure head in other approximate ways (Coulbeck 1984). In addition, using pump discharge as the decision variable in a lumped hydraulic model implicitly assumes there is a combination of pumps that supplies the optimal flow under the correct amount of pressure to cause the desired change in tank level. This assumption is increasingly difficult to satisfy as the network hydraulics become more complex in multisource and multi-tank systems.

In general, as the number of pumps or pump combinations increases, so does the computational advantage of the lumped pump-station parameter approach over the pump run-time approach. However, although the pump run-time approach yields the desired pump operational policy directly, the solution obtained using the lumped pump-station parameter approach subsequently must be translated into an appropriate pump policy. Although the computational time associated with this subproblem typically is a small fraction of the time required to solve the implicit control problem, it still is significant.

In general, the majority of optimal control algorithms were developed for applications with fixed-speed pumps. Variable-speed pumps simplify or exacerbate the difficulty of the problem, depending on the decision variable. If pump run time is chosen, each variable-speed pump is represented by a series of fixed-speed pumps. However, such a formulation increases the total number of decision variables and, hence, computation times. On the other hand, the wider continuous-range pump output of variable-speed pumps provides a better mechanism for implementing the continuous solutions associated with formulations of lumped pump-station parameters. Alternatively, pump speed is chosen as a continuous decision variable in the lumped-system formulation (Lansey and Zhong 1990).

Despite the multitude of control algorithms that were developed for optimal control of water supply pumping systems, several areas of potential research remain. For example, few researchers have investigated the development of optimal control policies for long-term (weekly) planning horizons. Similarly, little research has been conducted on the impact of final pump operations on pump maintenance requirements. Robustness of operations is also a neglected area. Finally, the design of water distribution systems is a well-examined area, but little emphasis is placed on the implications of design on operation and vice versa.

Although the use of expert-system technology or neural-network technology in either developing or implementing optimal control strategies seemingly has great potential, little work has been conducted in this area. Two applications of knowledge-based selection were described by Fallside (1988) and Lannuzel and Ortolano (1989). Fallside and Perry (1975) applied a decomposition approach to an existing system; however, after gaining experience and performing extensive systems analysis, they dropped the scheme in favor of a heuristic described as *pump priority logic* (Fallside 1988). Lannuzel and Ortolano (1989) also examined a water supply pumping system and developed an operational heuristic from experience. These rules of thumb were then combined with a simulation model in an expert system. Although both studies limited applicability to other systems, they nevertheless provide some insight into the usefulness of such an approach. Although several successful applications of optimal pumping control exist in Europe and Israel (Alla and Jarrige 1989; Orr and Coulbeck 1989; Orr et al. 1990; Zessler and Shamir 1989), widespread application of such technology in the US is severely limited. Future widespread applications of optimal control technology to domestic water supply systems are likely to depend on increased use of more sophisticated SCADA systems and the availability of more commercially available off-the-shelf control software. Additional problems to overcome include the necessity of well-calibrated network models and the availability of accurate demand-forecast models.

Such concerns highlight the need for systems analysts to work closely with operations personnel to develop and implement a particular control environment. In most cases, experienced operators already possess valuable insights into the operation of their system that proves to be crucial to the development of a successful control scheme. Ideally, the system analyst should work in concert with the system operator to develop an environment that the operator is not only comfortable with but feels some degree of "authorship" as well. In particular, the system should reflect the operator's existing wants and needs as much as possible while providing a framework for expanded control capabilities. In the final analysis, the real challenge of system analysis does not lie in the development of more sophisticated computer algorithms but in the development of more efficient strategies and programs for their implementation.

This page intentionally blank.

AWWA MANUAL M32

Appendix A

Hydraulic and Water Quality Modeling of Distribution Systems: What Are the Trends in the US and Canada?

ENGINEERING AND COMPUTER APPLICATIONS COMMITTEE REPORT

Co-authors: Jerry L. Anderson, PE, CH2M HILL, Louisville, Ky.
Mark V. Lowry, PE, Turner Collie & Braden Inc., Austin, Texas
James C. Thomte, PE, Bohannan-Huston Inc., Albuquerque, N.M.

Background and Utility Characteristics

In October 1999, the AWWA Engineering and Computer Applications Committee (ECAC) commissioned a survey of Water Network Modeling practices. The objectives of the survey were threefold, as listed below:

1. Provide up-to-date information for the revision of AWWA's M32—*Distribution Network Analysis for Water Utilities*

2. Gauge the current level of interest and activity in using computers to model water distribution systems, including the behavior of water quality, and identify potential trends.

3. Identify and share with AWWA members some interesting and possibly innovative, applications of water quality modeling in distribution systems.

The survey was distributed to 989 utilities in Canada and the United States. Utilities were selected via a review of AWWA's databases to select utilities serving populations greater than 35,000 people. Total number of responses was 174, representing an 18 percent return rate. The findings of the survey are contained in this report. Figure A-1 shows the number of utilities in each state and province that responded to the survey. The service population of respondents ranged from 21,500 to 8 million people. Figure A-2 shows a breakdown of utilities by service population. One third of the respondents serve between 50,000 and 100,000 people. Almost one-fourth serve between 100,000 and 200,000 people. The remaining size categories had fewer survey respondents. Approximately 68 percent of the respondents serve a population base of less than 200,000. Only 7 percent of the respondents served more than 1 million people. Several tables and graphs presented below show the results according to these service population categories to highlight differences, if any.

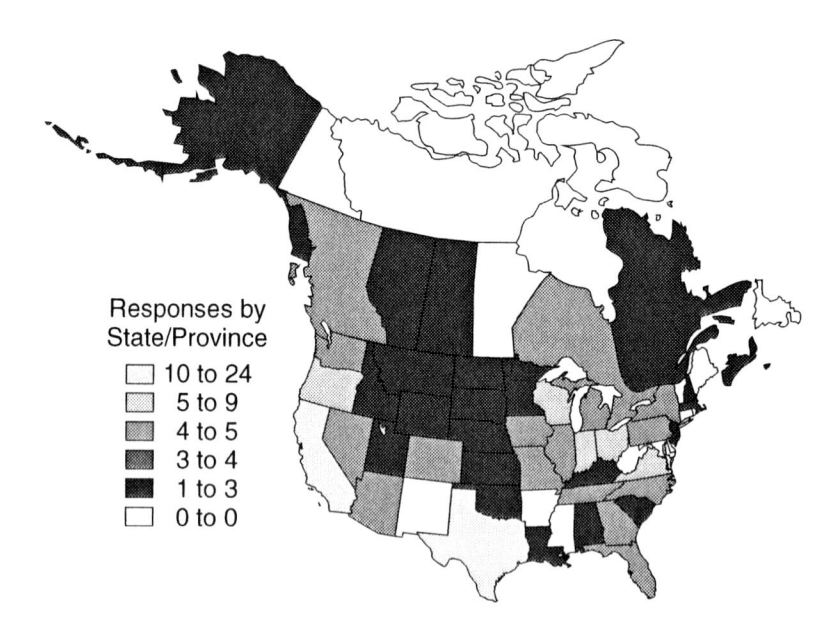

Figure A-1   Utility responses

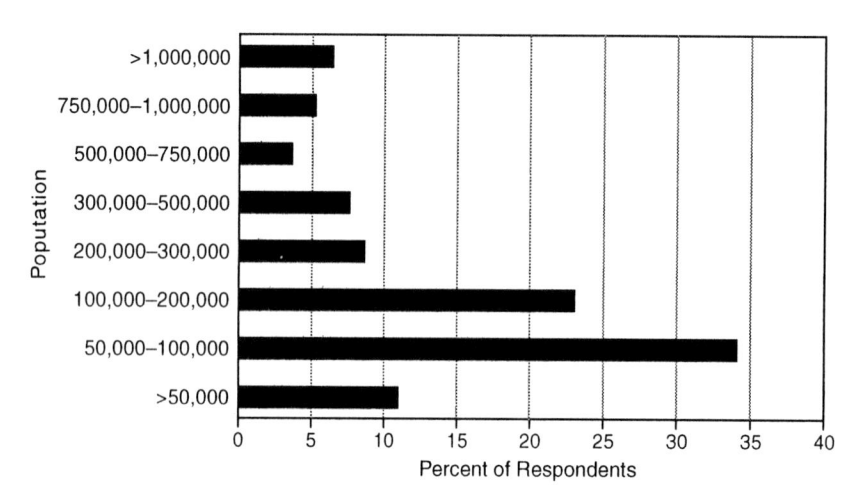

Figure A-2   Size of utility

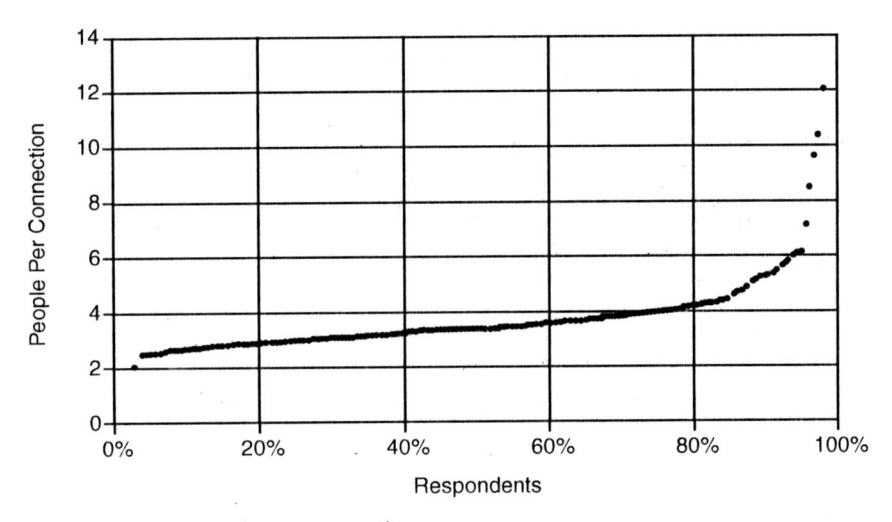

Figure A-3 Population per service connection

Service connections, or customers, were also identified. More than 80 percent of the respondents had 100,000 or fewer service connections. Only 8 utilities had greater than 250,000 service connections. Figure A-3 depicts the range of average service population per service connection, from 0.9 to 2,333 people per connection, ranked in ascending order. All but three respondents had 12 people or less per connection, so the top three were excluded from the figure. The very high values are a result of wholesale customers who purchase water through a few service connections but serve a large population. The median value is 3.4 people per connection. Average day demand (ADD) per utility ranged from 2.7 to 1,300 million gallons per day (mgd). The ranges of demands for respondents are shown in Table A-1. More than half the respondents (57 percent) had an ADD of less than 25 mgd, and three fourths of the respondents had an ADD of less than 50 mgd.

ADD is plotted in Figure A-4, ranked in ascending value of demand per person, and ranging from 75 to 584 gallons per day per capita (gpdpc). The high per capita values are unusual. The 90-percentile value is 275 gpdpc, which is less than half of the highest value. Climate is certainly one factor causing the high range of per capita demand, because many of the utilities above the 90-percentile are located in the southwest. However, climate is not the sole factor. Another cause for high demands would be utilities serving commercial and industrial customers that are large water consumers. Modelers who are developing demands based on per capita data must exercise caution and remember to compensate for nondomestic demands that could be a major component of total demands. The median per capital value (50-percentile) was 159 gallons per day per person.

Table A-1 Average day demand

Flow Rate, mgd	<10	10 to 25	25 to 50	50 to 100	100 to 250	250 to 500	>500	No Response
Number of Respondents	36	61	29	19	14	3	3	9
Percent of Respondents	21	36	17	11	8	2	2	NA

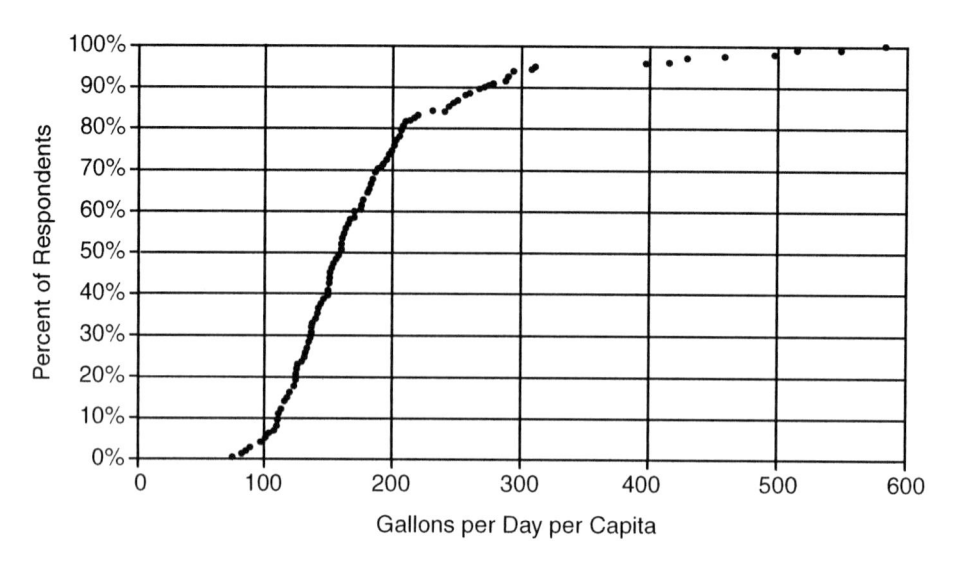

Figure A-4 Average per capita demands

Source Water Information

Respondents were asked to indicate the source of their drinking water. The responses showed a good cross section covering groundwater, reservoir, and river sources. Figure A-5 shows a breakdown of the respondents and types of source water. Forty-four percent indicated that more than one type of source water.

The maximum and minimum temperatures for the raw water sources were surveyed because of potential impact on water quality in the distribution system. For example, water temperature significantly affects the decay rate of chlorine residual. More than half of the respondents (57 percent) observe very high water temperature variations of between 30 and more than 60 degrees Fahrenheit in their water supply. Only 6 of the respondents (4 percent) have constant water temperatures. Figure A-6 shows the minimum water temperature and range of variation for all respondents ranked in order of ascending value of the range. The figure illustrates that water temperature varies for most utilities to the extent that it must be properly addressed when modeling water quality.

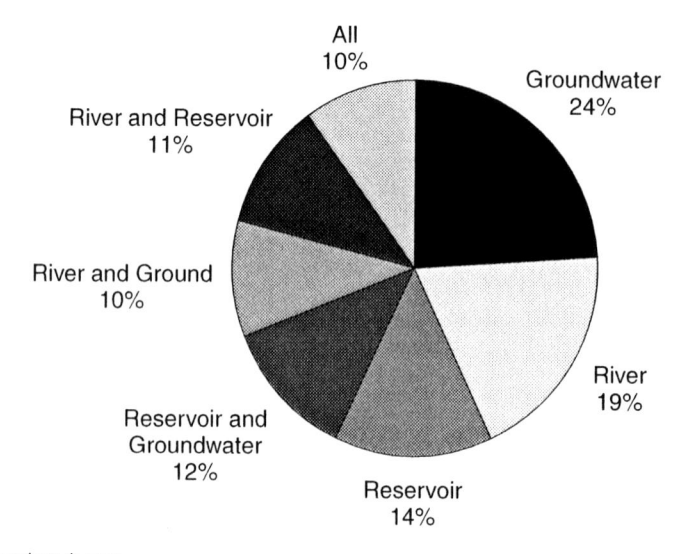

Figure A-5 Source water type

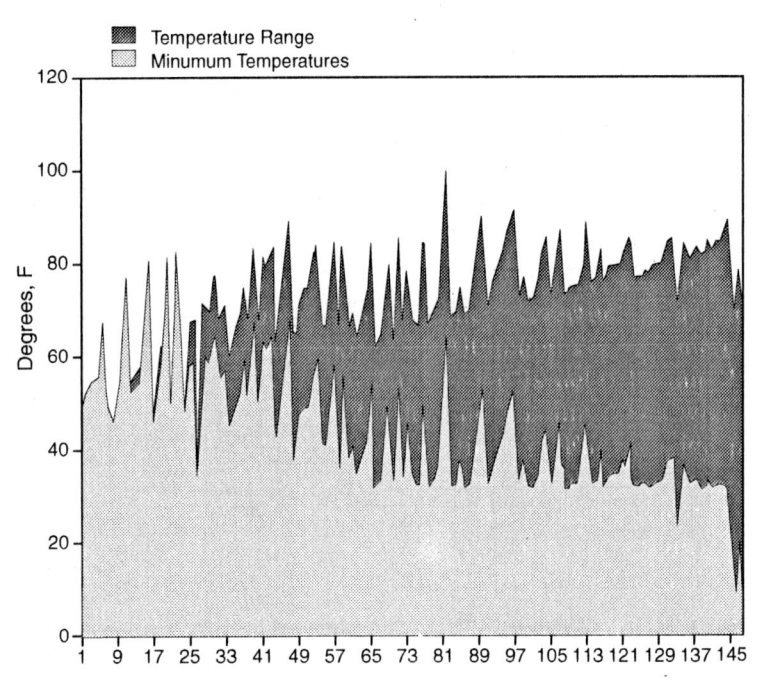

Figure A-6 Source water temperature

Distribution System Information

This section contains a series of questions directed toward the characteristics of the respondents' water distribution systems, describing piping, service pressures, storage tanks, and water quality issues; the questions often are about both current and planned conditions. The results are discussed below.

Pipelines

Respondents' distribution systems contained a range of 9 to 12,757 miles of pipeline. The majority of the respondents (69%) had distribution systems with less than 1,000 miles of pipeline, which is depicted graphically in Figure A-7. The median value is 460 miles. Figure A-8 shows the relationship between the population served and the number of pipelines in the distribution system. Because of some outlyers in the maximum and minimum values, the top and bottom three points were excluded from the graph. The median value is 270.5 people per mile. Sixty percent of the respondents ranged between 200 and 400 people per mile. These values may be helpful to a utility interested in estimating costs to prepare a geographical information system or distribution system model based on service population, if the length of pipeline is unknown.

The minimum pipe diameter in each system ranged from 1 to 14 inches. Three-fourths of the respondents indicated a minimum diameter of between 2 and 4 inches, inclusive. Figure A-9 shows the minimum and maximum pipe diameters for each utility arranged in ascending order of minimum diameter.

The maximum pipe diameter ranged from 14 to 700 inches; however, the top three maximum diameters were not plotted in Figure A-9 because they were much larger values than the maximums given by the other respondents and would inordinately skew the graph. Three fourths of the respondents indicated a maximum diameter of between 20 to 54 inches.

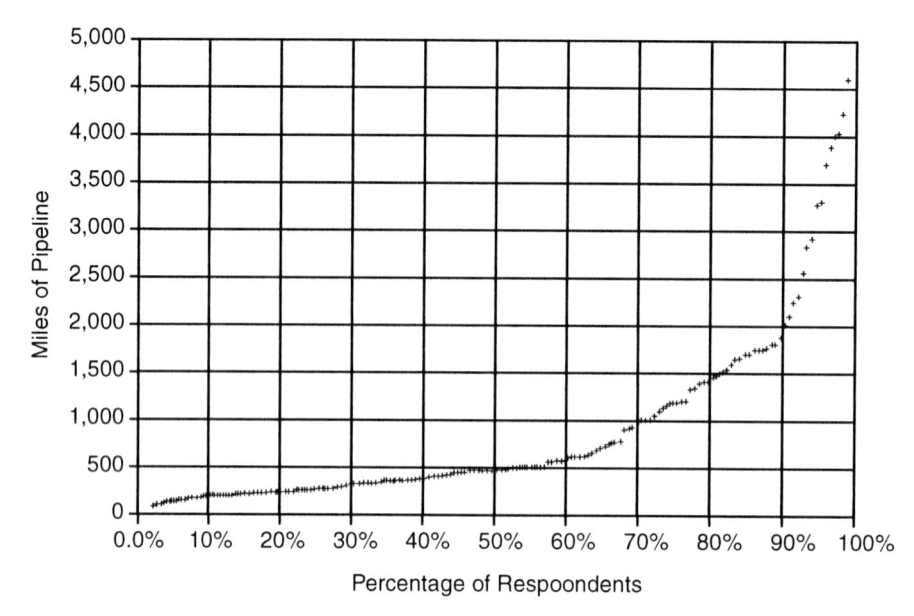

Figure A-7 Distribution system pipeline length

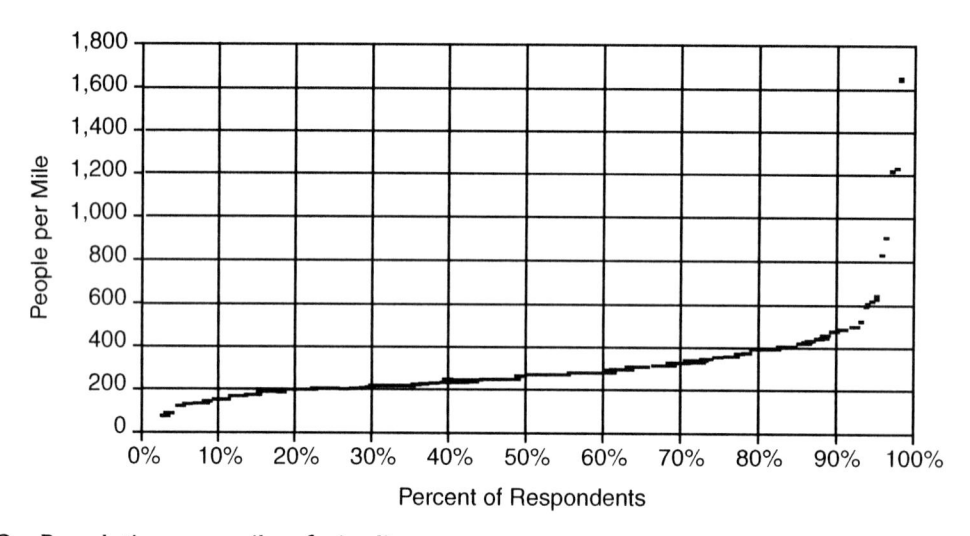

Figure A-8 Population per mile of pipeline

Pressure Zones

The number of pressure zones varied substantially from a single zone for 17 respondents to over 100 pressure zones for two respondents. More than half of the respondents had five or less pressure zones. Almost half of the respondents (46%) reported between 5 and 25 pressure zones. Table A-2 shows a breakdown of pressure zones.

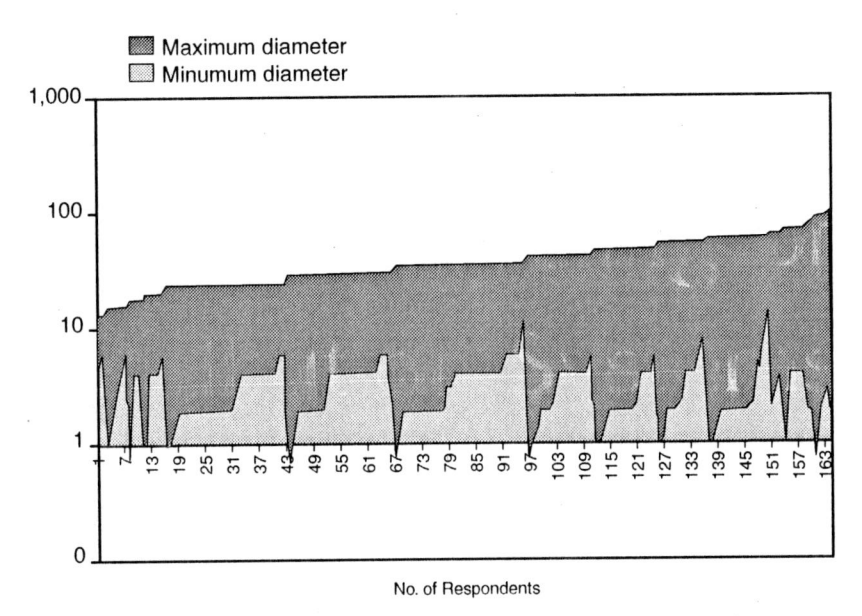

Figure A-9   Pipe diameters

Service Pressures

Figure A-10 shows both minimum and maximum service pressures for the respondents. The minimum service pressures ranged from less than 20 psi to a high of greater than 60 psi. One-third indicated a minimum pressure of between 30 and 40 psi. The maximum pressure ranged from a low value of less than 50 psi to a high value of greater than 200 psi. Ten of the respondents had maximum pressures exceeding 200 psi, which is unusually high. Some of the very high service pressures that were reported may be in transmission mains that do not directly serve customers. About half the respondents indicated their maximum pressure to be between 100 and 150 psi.

Tank Information

The surveyed queried the type and number of storage tanks in each system. The types were categorized as Fill/Draw and Inflow/Outflow Tanks. Fill/Draw tanks fill and draw in a distinct sequence, usually because they must be filled or emptied by pumping. On the other hand, Inflow/Outflow Tanks "float" on the system. The two types of tanks are set up differently in distribution system models and a modeler must be careful to properly simulate their operation. Most of the respondents indicated that they had both types of tanks in their distribution system but the inflow/outflow tanks were slightly more common, as shown in Table A-3.

Table A-2   Pressure zones

No. of Pressure Zones	<5	5 to 10	10 to 25	25 to 50	50 to 100	>100	No Response
Respondents	78	44	34	4	5	2	7
Percentage %	47	26	20	2	3	1	NA

Table A-3 Storage tanks

No. of Storage Tanks	None	<5	5 to 10	10 to 20	20 to 40	>40	No Response
Storage Tanks							
			Fill/Draw				
Respondents	32	48	46	20	9	1	18
Percentage %	21	31	29	13	6	1	NA
			Inflow/Outflow				
Respondents	35	48	27	14	13	5	32

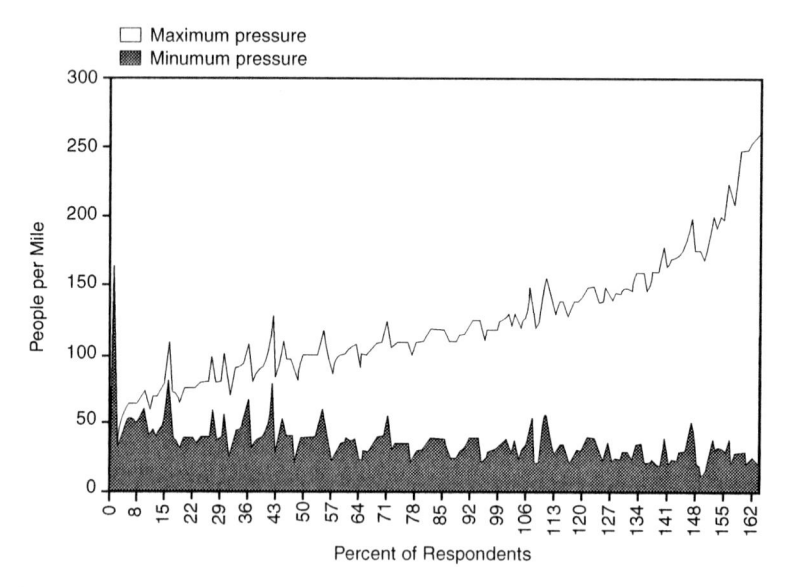

Figure A-10 Service pressure

Secondary Disinfectants

Secondary disinfectants are applied after treatment and remain as a residual in the distribution system. The most common secondary disinfectant, used by 89 of the respondents, was chlorine. The next most common was chloramine, used by 50 of the respondents, followed by chlorine dioxide, used by 6 of the respondents. Figure A-11 shows the breakdown of secondary disinfectants used by the respondents.

Water Quality Problems in the Distribution System

The purpose of obtaining information about water quality problems was to determine what the greatest needs were for water quality modeling. Several types of problems were identified:

- Loss of disinfection residual
- Taste and odor complaints
- High disinfection by-product (DBP) formation

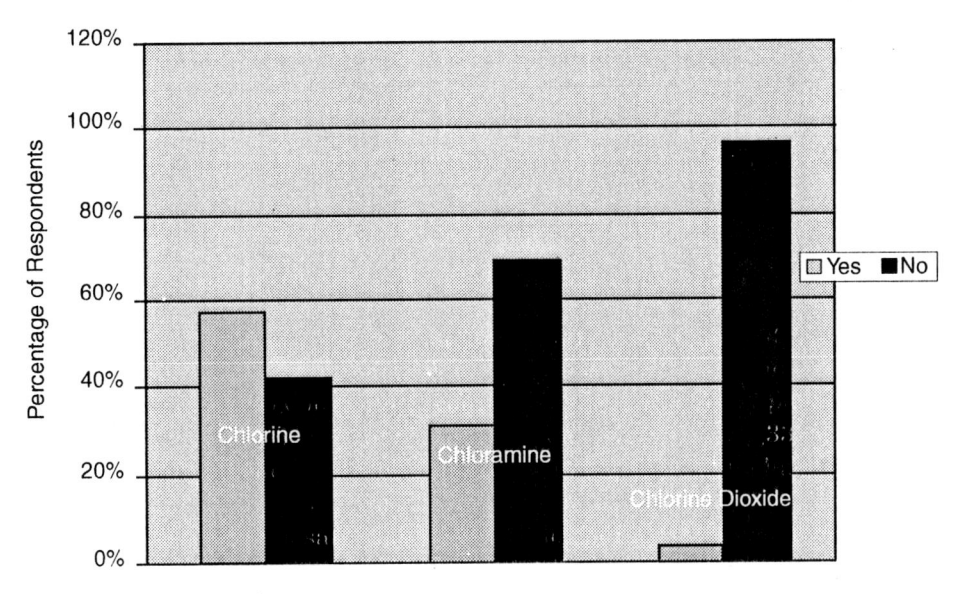

Figure A-11 Secondary disinfectants

- Excessive corrosion
- Coliform occurrences
- High heterotrophic plate counts
- Colored or turbid water

The responses are shown in Figure A-12 and categorized by frequency in Table A-4. The most common problems cited were taste and odor and turbid water. Only 15 out of 155 respondents indicated that they had no taste and odor problems and only 50 out of 154 respondents indicated no problems with turbid water.

The least common problems were high DBP formation and excessive corrosion. In both cases, 133 of the respondents indicated that these problems did not occur in their systems.

Hydraulic Modeling

This section contains a series of questions on the current and planned use of water hydraulic modeling. These questions relate the following general areas:

- Types and sizes of hydraulic models
- Sources of information for the models
- Calibration Information

The majority of respondents (150 of the 174) conducted some form of hydraulic modeling at their utility. Ten more utilities, representing an additional 6% of respondents, plan to conduct some form of hydraulic modeling in the future.

144 COMPUTER MODELING OF WATER DISTRIBUTION SYSTEMS

Table A-4 Type of water quality problem

	Number of Occurrences						
Type of Problem	None	1–5	5–10	1–20	20–50	>50	Total
Loss of Disinfection Residual	103	26	6	15	0	2	152
Taste and Odor Complaints	15	20	18	21	32	49	155
High Disinfection By-product (DPB) Formation	133	14	0	0	0	1	148
Excessive Corrosion	133	9	0	5	1	0	148
Coliform Occurrences	92	44	7	6	3	1	153
High Heterotrophic Plate Counts	112	12	4	5	4	7	144
Colored or Turbid Water	50	16	13	17	18	40	154

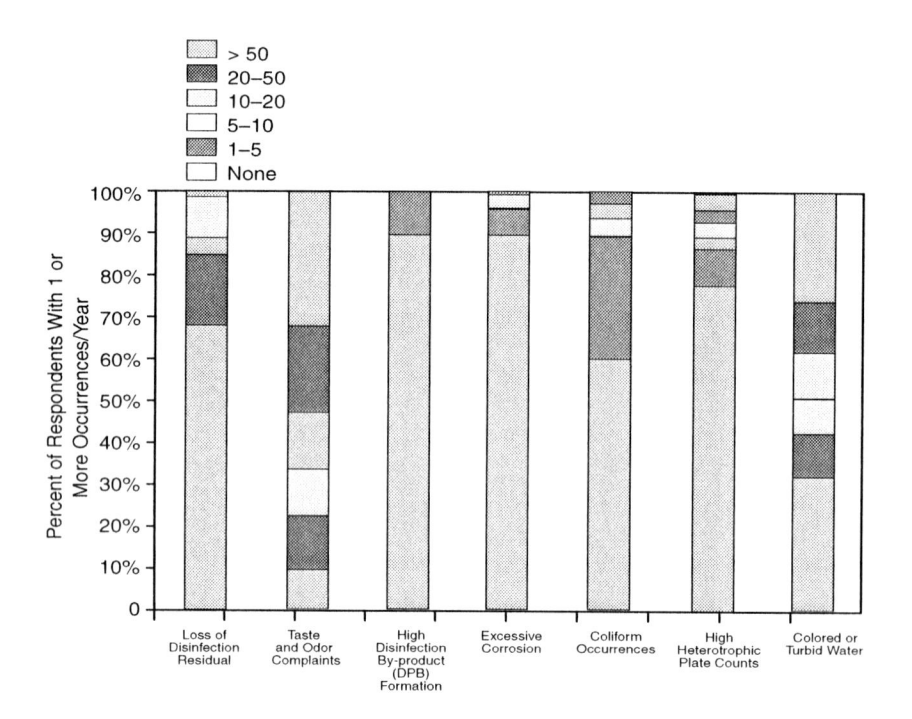

Figure A-12 Water quality problems

Model Information

A query was made to determine the current and planned types of modeling in use or being considered. The responses showed that most utilities have models of their transmission and secondary mains and predominantly do steady-state analyses of their distribution systems. Figure A-13 gives some indication of size of utility that answered the water hydraulic model questions. The size of the utility, as measured by its customer base, should be considered when reviewing the model information presented in this section. It is apparent from Figure A-13 that the majority of respondents (more than 50%) provide service to a population of 50,000 to 200,000.

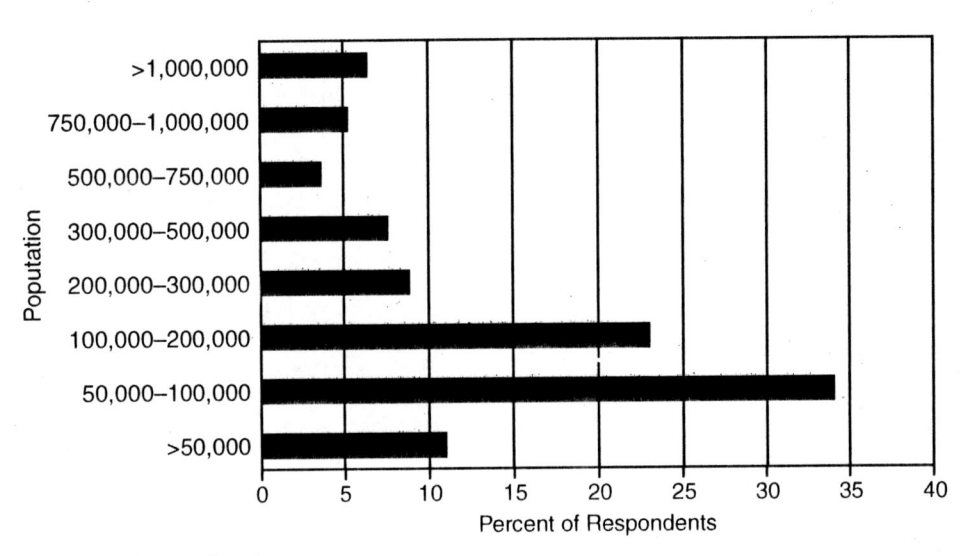

Figure A-13 Size of utility

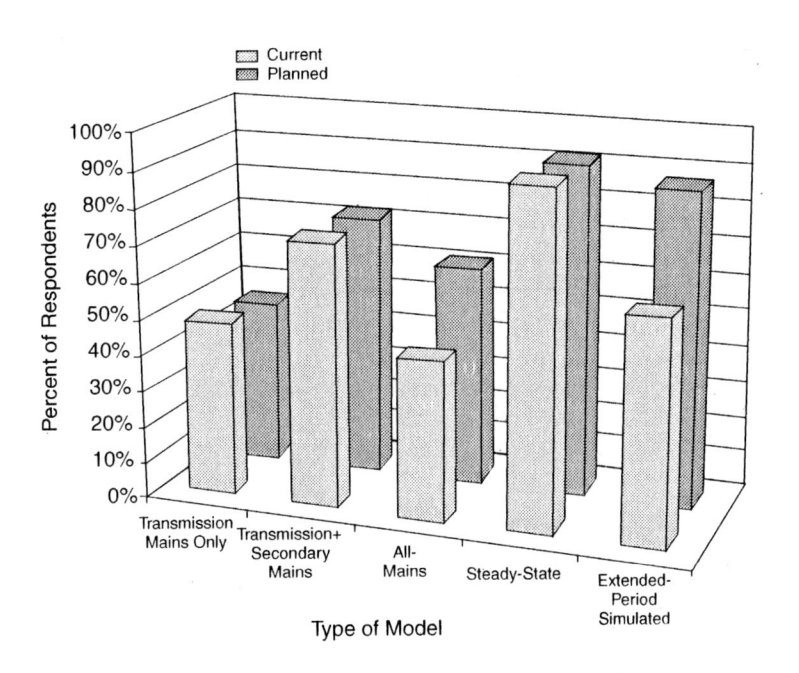

Figure A-14 Model types

Figure A-14 reflects the type of model used in terms of actual percentage of system modeled (skeletonized) and also time step, steady-state versus extended-period simulation (EPS), and the model the utilities use. Three categories represent the level of detail in the water system modeled, beginning with transmission lines (greatest skeletonization) to all mains, which represent all system piping but excluding service level piping. The responses to the question for planned use showed that utilities intend to maintain current uses but indicate an increase in the number of utilities planning to develop all-main models and the development of extended-time-period models. This indicates that utilities realize the need for more sophisticated models and are creating programs to provide that next level of sophistication.

Figure A-15 indicates current and planned model uses of the models by respondents. The most common use of a hydraulic model was in the area of master planning, which is generally a series of evaluations to determine the utility's future system requirements. Fire flow analyses, subdivision planning, and rehabilitation planning were nearly as popular as master planning. Energy management, used to minimize cost of providing service, was the least likely current use of a model, with only 27 of the utilities using this type of analysis. The planned use of the hydraulic models showed the continued use of the models for the traditional planning and fire flow analyses and showed a growing desire to use the model for energy management in the future. Clearly, utilities are recognizing the benefit in using the model for operational applications.

Respondents were asked to describe how frequently they use their model based on the various uses shown on Figure A-16, which illustrates frequency of use for daily, weekly, monthly, and yearly basis. Most utilities are using their models on either an annual or monthly basis. The most common daily use is fire flow evaluations, closely followed by subdivision expansion.

These indicate that models are used more by the segment of the utility responsible for assessing water availability for proposed development. These uses follow the more traditional planning activities. As the water model use is extended into the operations areas, those areas will likely experience more frequent model use, too.

Sources of Model Information

Respondents were asked what source they used for determining consumption and how they allocated that consumption to the model. Figure A-17 shows the current and planned sources of data and method of allocation used by percent of respondents. The most common source of consumption information was the utility billing system (117 of the respondents). Many utilities surveyed did not respond to these questions on consumption, illustrated in Figure A-17, especially under the future planned category. Of those that did respond, little difference between current and planned methodologies was noted.

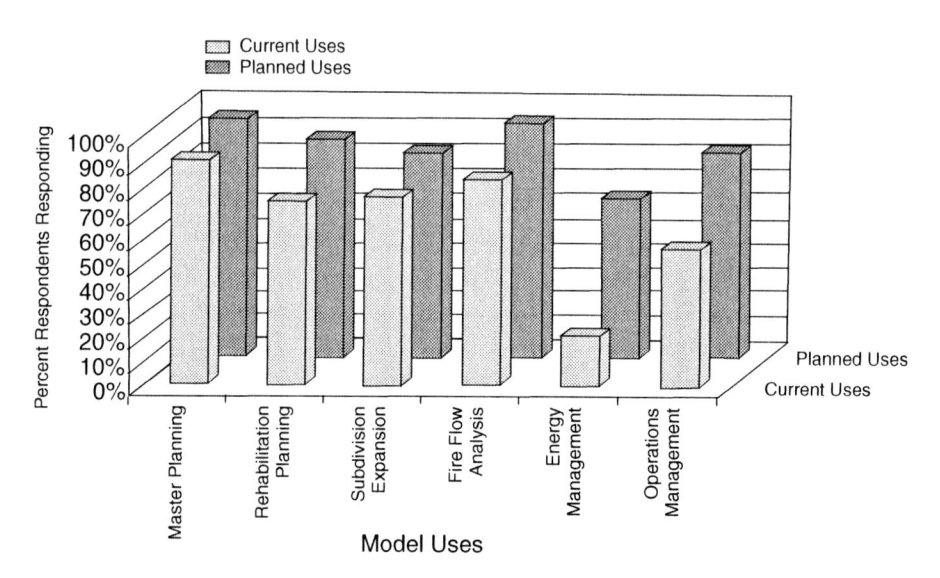

Figure A-15   Current and planned model uses

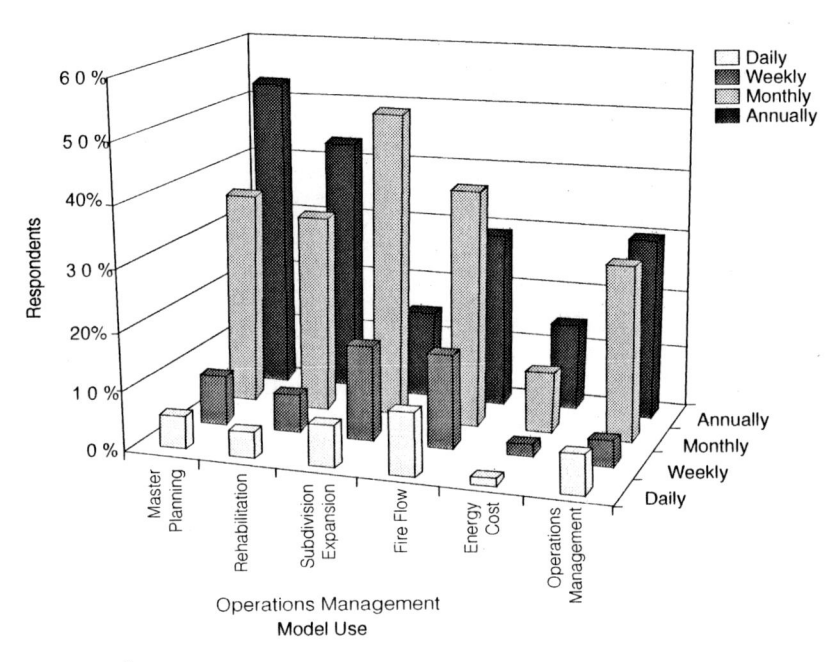

Figure A-16 Frequency of use

The types of field measurements conducted at each utility were surveyed. Pressure and flow were the most common, followed by consumption and C-factor testing. The least common type of field test was sodium sampling. Figure A-18 illustrates types of field measurements currently taken and those planned in the future. It is apparent, based on Figure A-18, that utilities are using traditional types of measurements, such as pressure, but are planning to expand their field measurements to include constituents, such as chlorine and sodium, for water quality modeling.

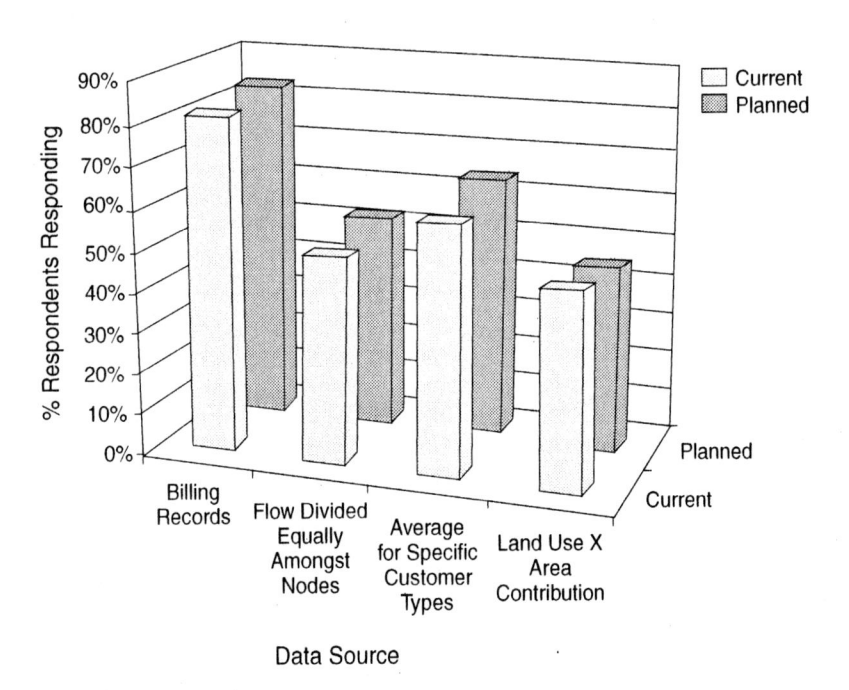

Figure A-17 System demand allocation method

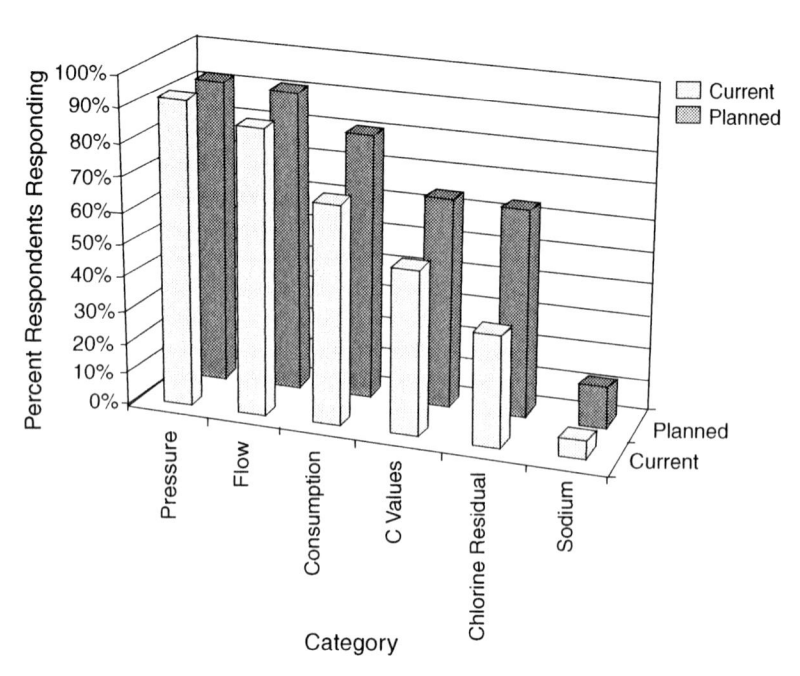

Figure A-18 Types of field measurements

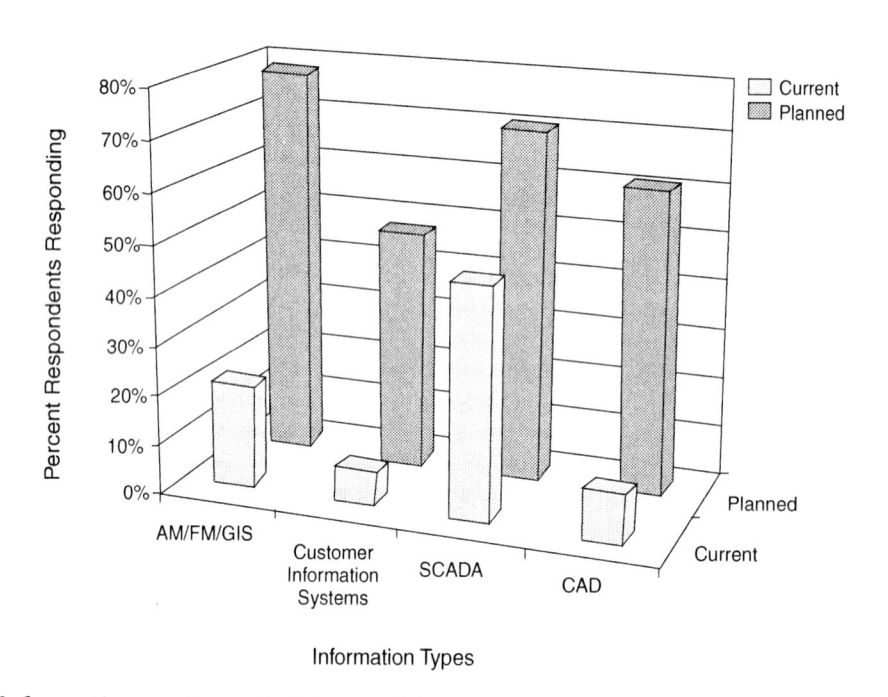

Figure A-19 Information systems tied to model

The respondents were asked to identify which information systems (IS) they linked their models or planed to in the future. More than any other category in the model portion of the questionnaire IS applications illustrate dramatic changes planned in the future. Figure A-19 shows the findings for IS applications including AM/FM/GIS and CAD for system configuration and geometry; Customer Information Systems for consumption data, customer complaint data; and SCADA for online calibration, operating scenarios, and energy management. The most common linkage is to a SCADA system (66 of the respondents). It is interesting to note, however, that in all areas considerable growth is expected in the future for development of linkages between information systems and models. The most dramatic anticipated growth is the desire to link to AM/FM/GIS, which increased from 29 to 109 responses. This information confirms the desire seen for an increased level of sophistication in distribution system modeling. As data from other sources are automatically loaded into the water model, the ability to evaluate model performance using real-time data can dramatically improve the model and its accuracy.

Calibration Information

Four questions were asked regarding model calibration:

- Who does/did the calibration of the models?
- When were the models last calibrated?
- What is the calibration accuracy?
- What is the calibration frequency?

The majority of the respondents indicated that they used consultants to complete their model calibration efforts and that the most recent calibration was completed within the last four years.

Figure A-20 illustrates the amount of respondents who regard their models as calibrated compared to system size. The percent of respondents that regard their own systems as calibrated ranged from a low of 78% to a high of 100%. However, 55% of the utilities did not respond to this question, possibly indicating that many of the respondents are unsure of the accuracy of their models. Of those responding, the actual definition of calibration accuracy varied. Some established an acceptable deviation from measured pressure (within 5 psi of measured, for example), while others based accuracy on a percentage of the measured pressure. Most of those who responded affirmatively on Figure A-20 typically used a calibration standard, such as, maintaining predicted pressure within 5% of actual system pressures. There seems to be no correlation between size of model or system versus the degree of calibration.

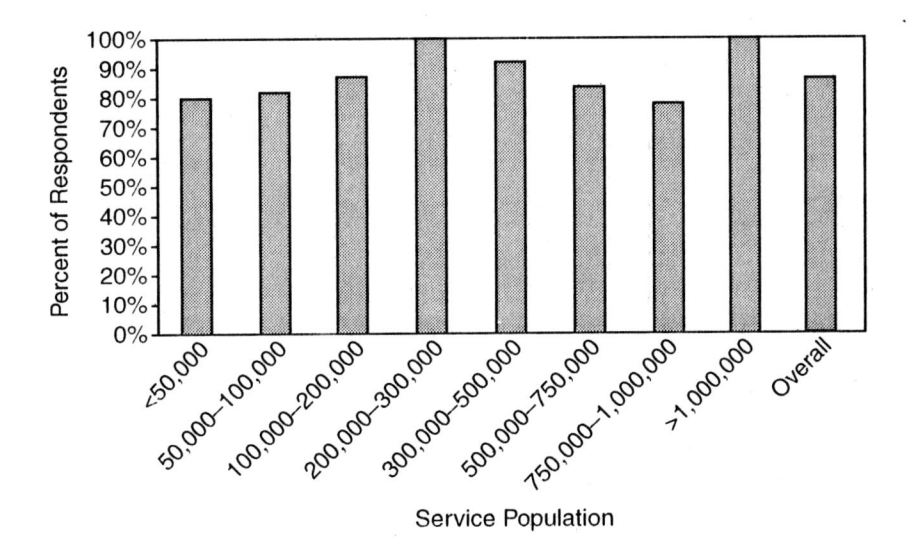

Figure A-20 Models that are calibrated

Similar to the accuracy of calibration response the most common response to the calibration frequency question was again no response (69 of the 174 respondents). Excluding those that did not respond, the majority of utilities indicated a calibration frequency of 2 to 5 years, or as needed.

Water Quality Modeling

This section contains a series of questions regarding the current and planned activities in water quality modeling. Based on an analysis of the survey results, 42 of the 173 respondents (24.3%) indicated that they were performing some form of water quality modeling. For those planning modeling in the future, the affirmative responses increased to 59%, indicating a substantial growing interest in water quality modeling. Survey respondents were queried for the following current and planned model characteristics.

Constituents modeled

The survey results show that chlorine and water age were the most commonly modeled parameters for both current and planned efforts. Both of these parameters provide insight into common issues related to water stagnation in the lines, and are related to the presence of tastes and odors in the system, and DBP formation. The results also show that there is significant growing interest in future modeling of trihalomethanes, haloacetic acids, and chloramines. The growing interest is likely driven by regulatory compliance issues and these models will provide a useful tool to evaluate scenarios for converting to new disinfectants or investigating DBP formation potentials. The responses received for this portion of the questionnaire are shown below in Table A-5.

Table A-5   Constituents modeled

Current No.	Respondents Replying Affirmatively			
	Current No.	Current %	Planned No.	Planned %
Chlorine	23	13.3	73	42.2
Chloramine	4	2.3	39	22.5
Fluoride	12	6.9	36	20.8
Phosphate	3	1.7	19	11.0
Sodium	4	2.3	14	8.1
Conductivity	5	2.9	23	13.3
Hardness/alkalinity	8	4.6	23	13.3
Temperature	5	2.9	26	15.0
Source blending	19	11.0	43	24.9
Water age	26	15.0	70	40.5
Trihalomethanes	7	4.0	42	24.3
Haloacetic acids	5	2.9	28	16.2
Other	1	0.6	33	19.1

Degree of Skeletonization

This indicates what fraction of total pipes are included in the model. For this survey an "all-main" model consists of all pipes or at least all pipe loops within the distribution system. The only pipes that would not be modeled in an all-main model are dead-end mains, which do not affect any circulation pattern and whose demand can be suitably represented at a node on the nearest pipe that is modeled. It is of particular concern in water quality modeling because of the need to ensure that all circulation patterns are included, that the percent contribution from each source can be positively identified, and that the smallest pipes with the potentially longest water age are modeled. The previously described Figure A-14 shows that 44% of who responded have all-main models. More important, the respondents planning to create all-main models in the future increased to 61%, demonstrating that the respondents see the value of all-main models for water quality modeling.

Steady-State Versus EPS

Another issue of interest concerns the use of steady-state versus EPS models. A steady-state model uses demand conditions and operational conditions for a single period or "snapshot" in time. Maximum day, maximum hour, minimum day and minimum hour are all common steady-state model conditions that are used. An extended-period simulation model is a series of snapshot models that have demand and operational conditions that simulate a time period, such as a 24-hr day, by conducting several steady-state analyses at equal time intervals, such as 1 hr. EPS requires operating data to simulate how the demand varies for each hour of the day and how the utility adjusts operational set points to accommodate greater or lesser demands. The survey indicates that EPS modeling by the respondents for water quality applications will be increasing in the future.

Purposes or Uses of Water Quality Models

Water quality modeling is currently being used for a wide range of purposes. Seeking operational information is the most common use, but other popular uses are to investigate water age and locate and size storage tanks. An example of operational information is to help design flushing programs for improving water quality. Several anticipated future uses were identified and listed below in Table A-6. Applications for which significant interest is shown for future use are optimizing disinfection residual, reducing water age, identifying stale-water zones, planning flushing programs, and improving operations.

Tank and Reservoir Modeling

Almost all current water quality models simulate storage tanks as completely mixed reactors. This is a simplification for convenience in most commercially available modeling software to avoid computational complexities and unknowns in modeling storage as something besides a completely mixed reactor. However, much research has been performed regarding the water quality in distribution system storage, demonstrating that storage water is far from completely mixed. Uniformity of water quality in storage is highly dependent on a number of factors including tank shape, inlet and outlet configuration, water temperature, ambient temperature, and average detention time. Therefore, the complete-mix assumption is an oversimplification. As a result, the use of multi-compartmented models or other advanced tank representation models will be more prevalent as more is learned about this subject. The questionnaire indicates that only two respondents are currently using a multi-compartment reactor model.

Table A-6 Water quality modeling applications

	Percent Responding Affirmatively	
	Current	Planned
Comply with regulations	11.0	34.1
Optimize residual disinfection	10.4	43.4
Optimize corrosion inhibition	3.5	24.3
Assess DBP exposure	4.6	28.3
Reduce water age	13.9	39.9
Identify/mitigate stale water zones	11.6	43.9
Plan/optimize flushing	8.1	45.7
Analyze pipe cleaning and replacement	8.1	34.1
Locate/size storage tanks	13.9	33.5
Improve operations	19.1	47.4
Other	0.6	2.3

Water Quality Calibration

Calibration of any model is a critical issue that determines the validity of the results. The degree of calibration and followup verification that is justifiable depends on the intended use. For example, if the purpose of a model is to only provide general guidance on operational issues, the degree of calibration may not be as important. On the other hand, using the model to help predict water quality constituent concentrations or to design new facilities would justify a high degree of calibration.

Table A-7 shows an equal number of respondents using historical data and intensive field investigations for water quality calibration. While historical data such as pressures, tank levels, and flow rates can be useful for hydraulic calibration, historical water quality data would be more challenging to use because of the difficulty in simulating historical water quality in an attempt to match the data. For the future, many more respondents are planning to perform some sort of water quality calibration. Of the 44 respondents who currently have a water quality model, eight did not calibrate at all, but this is expected to decline in the future, as shown in the table.

The ultimate degree of calibration would be demonstrated when the predictive results of the model could be reliably compared to the real time data from a Supervisory Control and Data Acquisition (SCADA) system. Water utilities that strive for this capability will find many useful applications for optimizing operations and managing water quality in the distribution system.

Table A-7 Method of water quality calibration

	No. Responding Affirmatively	
	Current	Planned
Historical data	18	49
Intensive field survey	18	36
None	8	4
Other	0	3

Index

NOTE: Note: *f.* indicates figure; *t.* indicates table; *n.* indicates note.

Analog models, 2
Area isolation, 6
AutoCAD, 2–3
Automatic calibration, 117
   analytical approaches, 117–118
   averaging simulation approach, 118
   bi-level computational framework, 119*f.*
   and expert system technology, 120
   genetic optimization methods, 119–120
   and GIS, 120
   gradient-based methods, 119
   heuristic methods, 119
   implicit and explicit bound constraints, 118–119
   implicit system costraints, 118
   macro-level, 120
   micro-level, 120
   optimization approaches, 118–120, 119*f.*
   pattern-search methods, 119
   problems in (underspecified), 121
   and SCADA, 120
   sequential simulation approach, 118
   See also Model calibration
Average day demand, 61, 65, 81, 137, 137*f.*
Average day simulations, 65
Average per capita demand, 137, 138*f.*
AWWA Engineering and Computer Applications Committee survey, 135–152
   utilities surveyed and statistical information, 136–137, 136*f.*, 137*f.*
   utility size, 144, 145*f.*

Booster stations and wells, 81, 94

C-factor tests, 46–47, 48
   gauge method, 47–48, 48*f.*
   parallel hose method, 47, 47*f.*
   See also Hazen-Williams formula
CADD. *See* Computer-aided design and drafting
Calibration. *See* Automatic calibration, Model calibration
Capital improvement programs, 3–4
Chlorine
   decay bottle test, 107–108, 108*f.*
   decay coefficients, 102, 108*f.*
   level monitoring, 6
   pipe wall demand coefficients, 102–103
   reactions, 99
   residual, 114, 115*f.*
   as secondary disinfectant, 142, 143*f.*
CIS. *See* Customer information systems
Completely mixed models, 123
Computer programs, 6
Computer-aided design and drafting, 9
   as data source, 17
   linked to models (survey), 148, 148*f.*
Conservation studies, 4
Continuity equations, 7
Cost equations, 7
Customer information systems, 2, 9
   and demand data, 29–30, 31–32
   linked to models (survey), 148, 148*f.*

Darcy-Weisbach formula, 23, 23*t.*, 24
Data, 7
   electronic records, 16–18
   management, 8
   paper records, 16
   from physical inspections, 18
   quality, 53–54
   sources, 15–18
   See also Demand data, Elevation data, Facilities data, Geographical data, Node data, Operations data, Pipeline data
Databases, 6, 11
Demand
   allocation, 11, 31–32
   average day, 137, 137*t.*
   average per capita, 137, 138*f.*
   fire flow, 61
   forecast models, 128
   maximum day, 61–62
   tank emptying and filling times, 63, 63*t.*, 64*f.*
   tank equalizing volume requirement, 63, 64*f.*
Demand data, 13, 14, 27
   adjusting demands, 32–34
   allocation, 11, 31–32
   in calibration, 56, 58
   commercial, 28
   from customer information systems, 29–30
   determining demand, 27–28
   determining limiting conditions, 60–61
   diurnal variations, 33–34, 34*f.*, 48–49, 49*f.*, 60–61, 60*f.*
   geographic variations, 32–33
   and GIS, 27
   industrial, 28

land use characteristics, 31
population counts, 30–31
residential, 28
seasonal variations, 33
sources, 29–31, 146, 147f.
water loss, 28–29
zone production data, 30
Disinfectants, secondary, 142, 143f.
Distribution system modeling, 1–2
benefits of, 3
database plus computer program, 6
in engineering design, 4–5
equation types, 7
frequency of use (survey), 11, 146, 147f.
hardware, 8
history, 2–3
and in-house vs. outside consultants, 9–10
model data, 7
model developer and decision maker, 10
model types (current and planned—survey), 144–146, 145f., 146f.
modelers' relationship with rest of utility, 10
modeling system components, 8
one-time vs. long-term use, 10
in planning, 3–4
preliminary considerations, 13
real-time, 11–12
and related software systems, 2, 8–9
software, 7
sophistication of, 11
survey on use of (1999), 10–11
in system troubleshooting, 5
in systems operations, 5–6
trends, 11–12
in water quality improvements, 6
See also Extended-period simulations, Steady-state simulations, Tank modeling, Water quality modeling
Diurnal demand measurements, 48–49, 49f.
example (residential, golf course, and total diurnal curves), 86–88, 87f., 88f., 88t.
in extended-period simulations, 85
in steady-state simulations, 60–61, 60f.
DTMs, 17
Dynamic programming, 129

Electronic maps, 32
Elevation data, 20–21
Emergency operations scenarios, 5
Energy costs
and extended-period simulations, 83, 90–91, 92f.
and optimal control systems, 130–131
and steady-state simulations, 71, 76–77
time-of-day rates, 83
Energy equations, 7

EPA. *See* US Environmental Protection Agency
EPANET, 2
EPS. *See* Extended-period simulations
Error reporting, 8
Eulerian approach, 101
Excel, 2–3
Extended-period simulations, 7, 55, 79–80
and alternate (e.g., natural gas) drivers, 91, 92f.
and booster stations and wells, 94
and control valve data, 81
and diurnal curves, 85
in energy optimization, 90–91, 92f.
establishing time interval, 80
in evaluating emergency system operations, 82
example (residential, golf course, and total diurnal curves), 86–88, 87f., 88f., 88t.
example (storage vs. production), 89, 89f., 90f.
Fullerton (California) operating-improvements case study, 93–94
input data, 80–82
model calibration process, 85–86
in operational improvements, 93–94
and other supply sources, 82
overall operational philosophy, 83
peak day philosophy, 83
peak week philosophy, 83
and pressure-regulating stations, 94
and pump curves, 81, 84
relation to steady-state simulations, 79, 80
reservoir or service storage data, 83
and reservoir requirements, 94
and SCADA information, 84
setup, 82–86
vs. steady-state simulations (survey results), 151
system operational data, 83
tank and reservoir data, 81
and time-of-day electric rates, 83, 90–91
verifying system operation, 80
vulnerability (reliability) analysis, 90
in water quality modeling, 100, 100t.
well data, 84

Facilities data, 13, 14, 21
pipe diameter, 22
pipe length, 22
pipe materials and properties, 22
pipe roughness factor, 23–24, 23t.
pump curves, 24–25, 25f.
pumps, 24
storage tanks and reservoirs, 26–27
valves, 25–26
Field tests and measurements, 103–104
laboratory kinetic studies, 107–108
planning, 42

problems in steady-state simulations, 59
tracer studies, 106–107
types used (survey), 147, 148*f.*
water quality surveys, 104–106
Fire flow
calculations, 8
demand, 61
and rehabilitating neighborhood distribution mains, 76
steady-state simulations, 65–66
studies, 4
tests, 52–53
Flow calculations, 5
Flow measurements, 42
hydrant pitot gauges, 43, 43*f.*
magnetic meters, 45
master meters (propeller meters), 44
pitot tubes, 44, 44*f.*, 45*f.*
problems with, 59
Fluoride, 99
Fullerton (California) extended-period simulations case study, 93–94

Geographic information systems, 2, 3, 9, 15
coordinates, 16
and demand data, 27, 31–32
and excess information, 16–17
linked to models (survey), 148–149, 148*f.*
in model calibration, 120
software, 2–3, 9
to track land use zoning, 31
Geographical data, 13, 14
coordinates, 16
and GIS, 15, 16
GIS. *See* Geographic information systems
Graphical user interfaces, 7–8, 11
GUIs. *See* Graphical user interfaces

Hardy-Cross method, 2
Hazen-Williams formula, 23–24, 23*t.*, 56, 57
Head loss measurements, 47
errors in should be less than head loss itself, 53–54, 121
gauge method, 47–48, 48*f.*
other methods, 48
parallel hose method, 47, 47*f.*
HGL. *See* Hydraulic grade line
Hydraulic grade line, 47, 48, 51
and data variations, 54
Hydraulic gradients, 59
tests, 51, 52*f.*
Hydraulic modeling, 126–127
calibration, 112, 113*f.*
to locate monitoring sites, 3
survey results, 143

Hydraulic tests and measurements, 41, 53

data quality, 53–54
planning field tests, 42
See also C-factor tests, Diurnal demand measurements, Flow measurements, Hydraulic gradient tests, Meter tests, *and under* Fire flow, Pumps
Hydrodynamic (computational fluid dynamics) models, 122, 122*f.*, 124

Information systems integration, 2, 8–9, 11

Lagrangian approach, 101
Linear programming, 129, 130

Mains
flushing programs, 6
rehabilitation programs, 4
steady-state simulations in rehabilitating neighborhood distribution mains, 76
Maintenance management systems, 17
Maps, 32
Mass balance models, 126
Maximum day, 61–62
Maximum hour, 62–65, 63*t.*, 64*f.*
Meter route maps, 32
Meter tests, 45
Metropolitan Water District of Southern California. *See* Fullerton (California) extended-period simulations case study
Microsoft Excel and Word, 2–3
MMS. *See* Maintenance management systems
Model calibration
directed-search, 109
in discovering system anomalies, 5–6
in extended-period simulations, 85–86
for flow, 56, 57*f.*
grouped, 109–110
for HGL, 56, 57*f.*
hydraulic, 112, 113*f.*
steady-state simulations in, 55, 56–60
survey results, 149–150, 149*f.*
in water quality modeling, 107, 109–110, 110*f.*, 112–115
See also Automatic calibration
Model development planning, 14
characterizing the system, 14
data sources, 15–18
degree of accuracy, 19
degree of detail, 18–19
elevation data, 20–21
skeletonization, 19
water system inventory, 14*t.*–15*t.*
Model formulation steps, 19, 20*f.*
Model setup, node data, 20
Network simulation models, 127

simplified, 127
Neural network models, 127–128
Node data, 20
Nonlinear programming, 129, 130
North Marin (California) Water District case study, 110
   background, 110–111
   chlorine residual, 114, 115f.
   hydraulic calibration, 112, 113f.
   sampling study, 111–112
   skeletonized representation of distribution system network, 111, 111f.
   Stafford Lake water, 112–113, 114f.
   water quality calibration, 112–115

Operations data, 13, 14, 35–36, 35t.
   from charts, 36
   dynamic, 34, 35
   from operations staff, 35
   SCADA system data translation/transfer/conversion issues, 38
   from SCADA systems, 36–37
   stable, 34
   from written records, 36
Optimal control problem, 129–130
   decision variables, 130
   objective function, 130–131
   operational constraints, 131
Optimal control systems, 125–126, 131–133
   and demand forecast models, 128
   and energy costs, 130–131
   heuristic approaches, 133
   and hydraulic network models, 126–127
   with lumped pump-station models, 132
   and mass balance models, 126
   and network simulation models, 127
   and neural network models, 127–128
   and optimal control models, optimal control problem, 129–131
   and optimization algorithms, 128–129
   and pump maintenance costs, 131
   and pump priority logic, 133
   with pump run-time models, 132
   and regression models, 126–127
   and SCADA systems, 125–126
   and simplified network models, 127
   and variable-speed pumps, 132
Optimization algorithms, 128–129

Peak day, 83
Peak week, 83
Personal computers, 11
Personnel training, 5
Physical scale models, 122, 122f., 123–125
Pipes and piping
   data, 21–24, 23t.
   miles of pipeline (survey), 139, 140f.
   pipe diameters (survey), 139, 141f.
   population per mile of pipeline (survey), 139, 140f.
   steady-state simulations in establishing system design criteria, 69
   traverse positions within, 44, 44f.
   velocity profiles, 44, 45f.
   and water quality modeling, 99
Pressure
   calculations, 5
   extended-period simulation in improved use of pressure-regulating stations, 94
   service, 141, 142f.
   steady-state simulations in establishing design criteria, 68
   zones, 140, 141t.
Pumps and pumping, 24
   bad assumptions based on nameplates, 59
   curve efficiency, 71, 71f.
   curves, 24–25, 25f., 56, 58, 70f., 81, 84
   and extended-period simulations, 81, 84
   sizing, 4–5
   steady-state simulations in energy savings criteria, 71
   steady-state simulations in establishing design criteria, 69–71, 70f., 71f.
   tests, 50–51, 50f.
   See also Optimal control systems

Regression models, 126–127
Reservoirs
   data, 26–27
   extended-period simulation in determining capacity requirements, 94
   and extended-period simulations, 81
   siting, 4
   sizing, 4
   See also Storage tanks
Roughness factor, 21, 23–24, 23t.

SCADA systems, 2, 9
   boundary condition data, 37
   communications components, 17
   and data for extended-period simulations, 84
   as data source, 17, 37
   and demand data, 32, 33
   human machine interfaces (HMIs), 17
   linked to models (survey), 148–149, 148f.
   master stations, 17
   in model calibration, 120
   and operations data, 36–38
   and optimal control systems, 125–126
   remote terminal units, 17
   verification or reference data, 37
Scenario generation, 8
Selective reporting of results, 8
Service pressure, 141, 142f.
Skeletonization, 19, 56, 58, 111, 111f.

survey results, 145*f.*, 151
Source water
　age tracking, 6
　load shifting, 5
　survey results, 138, 138*f.*, 139*f.*
Steady-state simulations, 7, 55
　average day simulations, 65
　in contingency planning, 77
　continuing use of, 77
　in developing system improvements, 56, 74–77
　emergency conditions and reliability, 67
　in energy optimization, 76–77
　in energy savings criteria, 71
　in establishing design criteria, 56, 67–74
　vs. extended-period simulations (survey results), 151
　and field measurement problems, 59
　fire flows, 65–66
　flow calibration, 56, 57*f.*
　HGL calibration, 56, 57*f.*
　and limiting demand conditions, 60–61, 60*f.*, 61*t.*
　in master planning, 75
　maximum day, 61–62
　maximum hour, 62–65, 63*t.*, 64*f.*
　in operator training, 77
　in outage planning, 76
　and piping system design criteria, 69
　plotting hydraulic gradients, 59
　and pressure design criteria, 68
　and pumping design criteria, 69–71, 70*f.*, 71*f.*
　in rehabilitating neighborhood distribution mains, 76
　relation to extended-period simulations, 79–80
　replenishment simulations, 66–67
　in selecting limiting conditions for design scenarios, 55, 60–67
　sensitivity analysis, 59–60
　simplifying initial simulations, 58–59
　in steady-state model calibration, 55, 56–60
　and storage facility design criteria, 72–74, 74*f.*
　stressing the system, 58
　in subdivision planning, 75–76
　testing assumptions, 56–58
　in water quality modeling, 100
Storage tanks
　data, 26–27
　elevated, 73
　emptying and filling times, 63, 63*t.*, 64*t.*, 64*f.*
　equalization, fire, and emergency, 72, 74*f.*
　equalizing volume requirement, 63, 64*f.*
　and extended-period simulations, 81
　number and location of in system, 73–74, 75*f.*
　service storage data, 83
　steady-state simulations in establishing design criteria, 72–74, 74*f.*
　storage allocation, 72–73, 74*f.*
　survey results, 141, 142*t.*
　and water quality modeling, 99
　See also Reservoirs, Tank modeling
Substance tracking, 6
Supervisory control and data acquisition systems. *See* SCADA systems
Systems models, 122, 122*f.*, 123, 123*f.*, 124

Tank modeling, 121–122, 125
　comparison of methods, 124–125
　compartment, 123
　completely mixed models, 123
　comprehensive, 123
　hydrodynamic (computational fluid dynamics), 122, 122*f.*, 124
　mathematical, 122, 122*f.*
　model types, 122, 122*f.*
　physical scale models, 122, 122*f.*, 123–125
　plug flow models, 123
　survey results, 151
　systems models, 122, 122*f.*, 123, 123*f.*, 124
TDH. *See* Total dynamic head
Tests and measurements, 41, 53
　data quality, 53–54
　planning field tests, 42
　problem areas, 59
　See also C-factor tests, Diurnal demand measurements, Fire flow tests, Flow measurements, Hydraulic gradient tests, Meter tests, *and under* Fire flow, Pumps
THMs. *See* Trihalomethanes
Time varying simulations. *See* Extended-period simulations
Total dynamic head, 50–51
Tracer studies, 106–107
Transport equations, 7
Trihalomethanes, 99
　formation coefficients, 103, 108*f.*
　formation test, 108

US Environmental Protection Agency
　and EPANET, 2
　and hydraulic modeling to locate monitoring sites, 3
USEPA. *See* US Environmental Protection Agency

Valves
　altitude, 26
　automatic control, 26

check, 25
and extended-period simulations, 81
flow control, 26
manual isolation, 25
pressure-reducing, 26
pressure-sustaining, 26
sizing, 4
vacuum-breaker, 26
Velocity head, 59

Water loss, 28–29
 calculations, 5
Water quality modeling, 6, 8, 11, 97
 advective transport of mass within pipes, 99
 applications (survey), 151, 152*t.*
 bulk decay coefficients, 102, 108*f.*
 calibration, 107, 109–110, 110*f.*, 152, 152*t.*
 chlorine decay coefficients, 102
 comparing modeled results to observed data, 109, 110*f.*
 computational methods, 100–101
 confirmation testing, 110
 data requirements, 101–103
 directed-search calibration, 109
 and Eulerian approach, 101
 extended-period models, 100
 governing principles, 99
 grouped calibration, 109–110
 hydraulic calibration, 112, 113*f.*
 hydraulic data, 101
 key problems for (survey), 142–143, 144*f.*, 144*t.*
 and Lagrangian approach, 101
 mixing of mass at pipe junctions, 99
 mixing of mass within storage tanks, 99
 need for, 97–98

North Marin (California) Water District case study, 110–115
 pipe wall demand coefficients, 102–103
 reaction rate data, 102–103
 reactions within pipes and storage tanks, 99
 set of equations in typical model, 99, 100*f.*
 steady-state models, 100
 THM formation coefficients, 103, 108*f.*
 use of (survey results), 150, 150*t.*
 uses of, 98–99
 water quality data, 101
Water quality monitoring sites, 3, 6
Water quality surveys, 104
 analysis procedures, 105
 calibration and review of analytical instruments, 106
 collection of ancillary data, 105
 communications and coordination, 106
 contingency plans, 106
 data recording, 105
 equipment and supply needs, 105
 logistical arrangements, 105
 personnel organization and schedule, 105
 preparation of data reports, 106
 preparation of sampling sites, 105
 safety issues, 105
 sample collection procedures, 105, 111–112
 sampling frequency, 104
 sampling locations, 104
 and system operation, 104–105
 training requirements, 105–106
Water system inventory, 14*t.*–15*t.*
Word, 2–3

Zone boundary selection, 5

AWWA Manuals

M1, *Principles of Water Rates, Fees, and Charges,* Fifth Edition, 2000, #30001PA

M2, Instrumentation and Control, Third Edition, 2001, #30002PA

M3, *Safety Practices for Water Utilities,* Sixth Edition, 2002, #30003PA

M4, *Water Fluoridation Principles and Practices,* Fifth Edition, 2004, #30004PA

M5, *Water Utility Management Practices,* First Edition, 1980, #30005PA

M6, *Water Meters—Selection, Installation, Testing, and Maintenance,* Fourth Edition, 1999, #30006PA

M7, *Problem Organisms in Water: Identification and Treatment,* Third Edition, 2004, #30007PA

M9, *Concrete Pressure Pipe,* Second Edition, 1995, #30009PA

M11, *Steel Pipe—A Guide for Design and Installation,* Fifth Edition, 2004, #30011PA

M12, *Simplified Procedures for Water Examination,* Third Edition, 2002, #30012PA

M14, *Recommended Practice for Backflow Prevention and Cross-Connection Control,* Third Edition, 2003, #30014PA

M17, *Installation, Field Testing, and Maintenance of Fire Hydrants,* Third Edition, 1989, #30017PA

M19, *Emergency Planning for Water Utility Management,* Fouth Edition, 2001, #30019PA

M21, *Groundwater,* Third Edition, 2003, #30021PA

M22, *Sizing Water Service Lines and Meters,* Second Edition, 2004, #30022PA

M23, *PVC Pipe—Design and Installation,* Second Edition, 2002, #30023PA

M24, *Dual Water Systems,* Second Edition, 1994, #30024PA

M25, *Flexible-Membrane Covers and Linings for Potable-Water Reservoirs,* Third Edition, 2000, #30025PA

M27, *External Corrosion Introduction to Chemistry and Control,* Second Edition, 2004, #30027PA

M28, *Rehabilitation of Water Mains,* Second Edition, 2001, #30028PA

M29, *Water Utility Capital Financing,* Second Edition, 1998, #30029PA

M30, *Precoat Filtration,* Second Edition, 1995, #30030PA

M31, *Distribution System Requirements for Fire Protection,* Third Edition, 1998, #30031PA

M32, *Distribution Network Analysis for Water Utilities,* Second Edition, 2005, #30032PA

M33, *Flowmeters in Water Supply,* First Edition, 1989, #30033PA

M36, *Water Audits and Leak Detection,* Second Edition, 1999, #30036PA

M37, *Operational Control of Coagulation and Filtration Processes,* Second Edition, 2000, #30037PA

M38, *Electrodialysis and Electrodialysis Reversal,* First Edition, 1995, #30038PA

M41, *Ductile-Iron Pipe and Fittings,* Second Edition, 2003, #30041PA

M42, *Steel Water-Storage Tanks,* First Edition, 1998, #30042PA

M44, *Distribution Valves: Selection, Installation, Field Testing, and Maintenance,* First Edition, 1996, #30044PA

M45, *Fiberglass Pipe Design,* First Edition, 1996, #30045PA

M46, *Reverse Osmosis and Nanofiltration,* First Edition, 1999, #30046PA

M47, *Construction Contract Administration,* First Edition, 1996, #30047PA

M48, *Waterborne Pathogens,* First Edition, 1999, #30048PA

M49, *Butterfly Valves: Torque, Head Loss, and Cavitation Analysis* First Edition, 2001, #30049PA

M50, *Water Resources Planning,* First Edition, 2001, #30050PA

M51, *Air-release, Air/Vacuum and Combination Air Valves,* First Edition, 2001, #30051PA

M54, Developing Rates for Small Systems, First Edition, 2004, #30054PA

To order any of these manuals or other AWWA publications, call the Bookstore toll-free at 1-(800)-926-7337.

This page intentionally blank.

Printed in the United States
49925LVS00003B/43-54